Public Safety Networks from LTE to 5G

Public Safety Networks from LTE to 5G

Abdulrahman Yarali
Murray State University
Murray, KY, USA

This edition first published 2020
© 2020 John Wiley & Sons Ltd

The right of Abdulrahman Yarali to be identified as the author of this work has been asserted in accordance with law.

Registered Offices
John Wiley & Sons, Inc., 111 River Street, Hoboken, NJ 07030, USA
John Wiley & Sons Ltd, The Atrium, Southern Gate, Chichester, West Sussex, PO19 8SQ, UK

Editorial Office
The Atrium, Southern Gate, Chichester, West Sussex, PO19 8SQ, UK

For details of our global editorial offices, customer services, and more information about Wiley products visit us at www.wiley.com.

Wiley also publishes its books in a variety of electronic formats and by print-on-demand. Some content that appears in standard print versions of this book may not be available in other formats.

Library of Congress Cataloging-in-Publication Data

Names: Yarali, Abdulrahman, author.
Title: Public safety networks from LTE to 5G / Abdulrahman Yarali, Murray
    State University, Murray, KY, USA.
Other titles: Public safety networks from LTE to Five G
Description: Hoboken, NJ, USA : John Wiley & sons Ltd., 2019. | Includes
    bibliographical references and index.
Identifiers: LCCN 2019030072 (print) | LCCN 2019030073 (ebook) | ISBN
    9781119579892 (cloth) | ISBN 9781119579908 (adobe pdf) | ISBN
    9781119580133 (epub)
Subjects: LCSH: Emergency communication systems. | Public safety radio
    service. | Mobile communication systems.
Classification: LCC TK6570.P8 Y37 2019 (print) | LCC TK6570.P8 (ebook) |
    DDC 384.6/4–dc23
LC record available at https://lccn.loc.gov/2019030072
LC ebook record available at https://lccn.loc.gov/2019030073

Cover Design: Wiley
Cover Images: © Photostriker/Shutterstock, © resilva/Shutterstock,
© Mr.Suchat/Shutterstock, © streetlucifer/Shutterstock

Set in 10/12pt WarnockPro by SPi Global, Chennai, India
Printed and bound in Singapore by Markono Print Media Pte Ltd

10 9 8 7 6 5 4 3 2 1

This book is dedicated to my kids, Fatemeh Zahra and Sadrodin Ali.

Contents

Preface

The public safety community has undertaken significant strides towards strengthening its abilities and capacity, and improving the communication of emergencies. Public safety depends on fast and efficient levels of communication in order to properly relay time-sensitive and critical pieces of information. The first responders, however, are still limited due to the presence of fragmented networks and the use of decades-old technologies in public safety agencies. They are using land mobile radio (LMR) for the majority of their communications and, as a result, public security has struggled to communicate across jurisdictional and company lines.

The Department of Homeland Security's Office of Emergency Communications in the United States of America developed a document with the complete collaboration of the National Council of Statewide Interoperability coordination and SAFECOM. The document is designed to address a few concerns about community safety/public safety. According to this document, they will provide help to educate the selected appointed officials and the public safety community about future emergency communications. The document also includes the evolution of emergency communications and their relevancy to the traditional land mobile radio communications with wireless broadband in the future, and the requirements to achieve the desired long-term convergence state. There are five critical success elements that are outlined for SAFECOM interoperability, namely technology, training and exercise, governance, standard operating procedures, and usage. The community's vision is to evolve public safety communications.

The deployment of a cost-effective wireless broadband that is aligned with Long-Term Evolution (LTE) to help keep up with the advancements in technology is currently underway to carry high speed transmission, which can transfer real-time information such as voice, images, data, and videos with high quality and high speed to guarantee connectivity anywhere, anytime, and for everyone for ubiquitous communication and collaboration between rescue workers. Different government bodies are working on improving public communication through measures that support the advancement of features within the 4G LTE network and the development of the next stage generation that will be able to support public safety communications.

Broadband has become the future of critical communication through its powerful innovative solution for better protection of public safety. It allows the delivery of real-time information, reliability, and performance of the core goal found in mission critical technology. Combined with other network elements and new features due of the advancement of technology, LTE has transformed public safety communication. The improvements and features added to this technology have paved the way to innovative solutions towards the next generation of technology that can support the increasing need for bandwidth within the frequencies.

The next generation technology consists of a collection of technologies, and utilizing the already existing fixed networks in place for wireless connectivity will be able to help cater to the different elements, including that of the spectrum, high performance, and stable infrastructure to offer emergency responders services. Network densification is the main element that has driven public safety towards the building and implementation of 5G technology within the public safety system. The linkage between 4G–5G evolution and spectrum for public safety is viewed as per the worldwide and European controllers, for example, ITU1, ECC, and the EC, range serving Open Safety (PPDR)2 activities must be found in the 700 MHz band, yet it is up to every nation to decide how, and how much.

One of the main reasons why 5G will be offering better services than 4G is the Internet of Things. The Internet of Things refers to the processes, people, and data that come together to make connections in the network that are more relevant and valuable. As the digital industry moves towards a more heterogeneous technology that contains a multi-layer network that consists of a microcell, low powered small cells, and relays supported through the digital connectivity, 5G is the most technically feasible technology that is able to meet these needs in public safety. A 5G execution will likewise incorporate all of the 4G usefulness, and all required public safety administrations will be accessible. Nevertheless, sitting tight for the next innovation level or 3GPP release, since it is proposing better administration for public safety, can undoubtedly trigger an everlasting hold-up circle while the following innovation level is continually offering something more and is superior to the last one.

Creation of a single cohesive wireless network for specific use by all the public safety and response teams would enable ad hoc information sharing without barriers to communication. It would also allow the sharing of physical and virtual resources across agents through the development of heterogeneous devices. A wireless network of information sharing eliminates the need for specific devices in communication. It increases trust among stakeholders in emergency response and public safety. The members of different groups would have unlimited access to information shared across networks. Shared networks enhance interoperability by allowing control over available resources thus tackling the issues in information and resource sharing.

This book will discuss the evolution of public safety system requirements by providing a technical analysis of the existing public safety network, 4G LTE and its applications in public safety and security, a comprehensive analysis of 5G network technology, the link between the most recent application of 4G and 5G, efficient utilization, and the spectrum sharing for public safety communications systems.

Acknowledgment

I would like to express my gratitude to all those who provided support and discussions, talked things over, read, wrote, offered comments, allowed me to quote their remarks, and assisted in the editing and proofreading. I would like to give special thanks to all my graduate and undergraduate students in TSM320, TSM322, TSM323, TSM397, TSM411, TSM421, TSM571, and TSM610 classes of our distinction program of Telecommunications Systems Management at Murray State University, Kentucky. This book would never have found its way to the publisher without these students.

1

Public Safety Networks from TETRA to Commercial Cellular Networks

1.1 Introduction

The telecommunications industry has been around for several decades and the changes that have occurred have completely changed how the world operates. In fact, because of the telecoms industry, the entire globe is now connected in one way or another. Legacy and traditional telecoms has also seen lots of "band-aid" technologies in which legacy equipment is combined with more legacy equipment in order to achieve new contracts or investments. That time is mostly over and the theory is that in the next three years, the entire telecoms industry will switch to more modern and agile platforms. These changes will revolve around and include 4G, artificial intelligence, machine learning, augmented reality, virtual reality, 5G, and cross-industry alliances. Public safety depends on fast and efficient levels of communication in order to properly relay time-sensitive and critical pieces of information, which could mean life or death. Unfortunately, many first responders today still rely on fragmented networks and decades-old wireless tech. They are using land mobile radio (LMR) for the majority of their communications, but the need for a faster, more reliable cellular/mobile communication has arrived. Wireless broadband will most likely be the technology to meet that need.

The Public Safety Group made considerable advances toward fortifying national preparedness and enhancing emergency communications abilities. In the current world we live in, the majority of public safety organizations in the world utilize dedicated communication systems such as terrestrial trunk radio (TETRA), TETRAPOL (a highly spectrum efficient frequency division multiple access [FDMA] technology), and Project 25. These are systems that were discovered more than two decades ago and their primary design was to offer highly reliable and secure mission critical narrowband voice-centric services to meet the requirements of the public safety communication users. Some of the specialized services provided by these systems include group and priority calls with push-to-talk (PTT) future and device-to-device communication, among others (TCCA 2018). However, advancements in public safety functions have been challenging over the years. At the moment, the majority of governments, research institutions, and public safety agencies are looking into ways of enhancing communication potential. The voice services at the moment can be considered to be at a satisfactory level because the data transmission capabilities are limited. In the age we live in, having a satisfactory level can be ineffective because of the various telecommunication advancements. This means that the concentration on voice-centric services has promoted a situation in which technology that is used for public safety communication is underdeveloped as compared to the

Public Safety Networks from LTE to 5G, First Edition. Abdulrahman Yarali.

technology used in the commercial domain in regards to available data rates. In addition, public safety systems are so fragile that they are affecting particular technical decisions that need to be made to ensure connectivity for the users anywhere and anytime. The study shows that the new Long Term Evolution (LTE) functionalities that are in the process of being designed for public safety communications have the potential of replacing TETRA by providing similar functionalities, but at an advanced stage. These new implementations will mean that the LTE architecture will have to get more network entities, especially application servers, because they will assume this role. According to research done by Stojkovic (2016) "the future LTE systems services will match or surpass those of TETRA and ensure they meet the public safety user requirements for the group and device-to-device communication by group call system enablers for LTE (GCSE-LTE), and proximity services (ProSe) functionalities. Moreover, the PTT service can be utilized to assist in mission-critical push-to-talk (MCPTT) in the new technology." Some of the approaches discussed in this chapter that will take the public safety networks (PSNs) to a whole new level include softwarization and virtualization as defined in software defined networking (SDN). These software networks assist in defining the clear separation between the data and control plane, meaning that there will be a quicker convergence in the newly installed networks, a centralization of policy decisions, and flexibility in reconfiguring network functions whenever there is overload or hardware damage.

PSNs refer to dedicated telecommunication networks used by public safety agencies, for instance, police, fire, emergency, and many more for critical communications (Ferrãos and Sallent 2015a,b,c). In the contemporary world, most public safety organizations use dedicated systems such as TETRA, ARCP Project 25 (P25), and TETRAPOL, all using narrowband technology (Chavez et al. 2015). The design for these systems is built to offer highly reliability and secure narrowband services.

Public service networks often have a very critical objective, especially with human lives depending on the successful exchange of information such as devices location, dispatch services, or alert messages. Fire brigades or police departments are organized in highly mobile groups of individuals respecting a strong hierarchical configuration, which necessitates different levels of priorities and end-to-end security. They may work in very harsh environments such as earthquake destruction, fires, or flooding, and are expected to provide relief under any circumstances and in any location. Beyond the voice and specific services that they receive from their rugged mobile devices, PSN users have tended to start taking advantage of more evolved and powerful applications, such as real-time video, temperature or heart-rate sensing from their private commercial smartphones (Favraud and Nikaein 2015).

With the consideration that even teenagers own a device much more powerful than first responders, several countries and specialized organizations have launched the definition or funding of the evolution of the PSN. Here, their main aim is to integrate the latest advances of mobile communications and offer technologies at a level equivalent to public commercial networks to their users. For example, in the UK, the Emergency Services Network (ESN) was set to start its deployment in 2016 and their main targets were fire-fighters and ambulances (NOKIA LTE 2014).

On the other hand, in the US, the FirstNet dedicated LTE network was set to be interoperable, highly reliable and resilient wireless broadband networks to serve police, fire-fighters, paramedics, and command centers. The TETRA and Critical

Communications Association (TCCA) recently selected LTE as the technology to complement TETRA for broadband communications (First Responder Network Authority 2012). Enhancements to LTE and LTE-Advanced (LTE-A) have been designed and standardized under Release 12 of the 3rd Generation Partnership Project (3GPP), as described by Doumi et al. (2013), to cope with the additional features necessary to support PSN operations. At the same time, various new technologies aiming to enhance the performance and efficiency of digital communication networks have sprung up in the past years, such as SDN (Wetterwald et al. 2016). It introduces a clear decomposition of control and data plane, which permits one to perform forwarding at flow level according to fine-grained rules. Secondly, there is also network function virtualization (NFV) and service function chaining (SFC) that complement SDN by adding further flexibility and increasing scalability. Hence it is a natural tendency to propose the benefits of these new technologies to the design of future PSNs.

In this chapter will first look at the challenges PSNs face by developing solutions for data rates such as introducing broadband data services into safer communication. Secondly, there will be a discussion on the integration of recent technologies, for example, cellular LTE and how such networks can be upgraded using SDN and NFV. This softwarization will be part of the solutions to the issues raised with commercial LTE and how PSN communications can be enhanced. In this chapter, the potential of these technologies is associated with a list of important characteristics such as rapid deployment, reliability, security, and dynamic resource management.

1.2 Evaluation of TETRA and TETRAPOL

Developed by the European Telecommunications Standardization Institute (ETSI), this is an open telecommunication standard for a public safety telecommunication system. This system has been in existence since 1995 and is deployed in countries such as Europe, Middle East, Asia, and Latin America. TETRA is known to mostly provide voice and dispatch services to mobile terminals functioning in direct mode operation (DMO) or trunked mode operation (TMO) through a dedicated infrastructure, or alternatively using the neighboring terminals as relays. On the other hand, data services are also available but they exist at very low speeds, from 80 kbps to probably the highest being 500 kbps, while using the TETRA Release 2, also referred to as the TETRA enhanced data service (TEDS). The big geographical area covered that is enabled by the use of low-frequency bands allows one to set up an infrastructure requiring few base stations. Closely related to this type of TETRA is the collaborative Project 25, which standardizes an equal system for LMR and has been greatly used in North America, Australia, and New Zealand. The latest version of this system defines data services at a lower level of data rates, which is quite similar to TEDS.

Furthermore, the TETRA standard can be perceived to be a suite for other standards. When it comes to practice, these standards take care of different technical aspects such as the air interfaces and network interfaces as well as services and facilities. As a worldwide standard, as discussed earlier, the majority of TETRA networks exist in many cities in the world. It is undeniable that these systems come with advantages. For instance, one advantage of TETRA is its interoperability certification requirement. This means that the system can be used and shared by all public organizations thereby reducing financial expenditure. Secondly, there is a higher guarantee for communication security during emergency and disaster situations. Being a fully digital system, TETRA ensures high-quality voice transfer and a wide range of data transfers. Additionally, TETRA gives support to voice-, circuit-, and packet-switched data transmission with varying bit rates and error correction levels.

According to Stojkovic (2016), a TETRA system is built on virtual private network technology and exists on one physical network that can be shared among several organizations. This means that each user group is able to use a TETRA based network because of its availability to the group. During disaster situations, the nature of a crisis often has an effect on the usable media. Let us take an example of a sudden stadium attack, which most of the time makes the public cellular technology useless because everyone will be on their mobile phones under a limited number of base stations.

Although we have seen the advantages of the TETRA based network, it also comes with disadvantages, and the major one is that it has a huge limit to data capacity. In the course of this chapter, the challenge will be comprehensively discussed as well as the possible solutions that can be put in place to meet PSN expectations.

Both trunked national radio systems of TETRA and TETRAPOL offer services like digital encrypted voice transmission and data transmission (connection oriented and connectionless). There are many voice services available for these radio platforms such as point-to-point (PtP) individual calls, Group point-to-multipoint (PMP) calls, emergency calls, and broadcast calls. TETRAPOL is a very suitable standard for mission-critical public safety users. TETRA is a time division multiple access (TDMA) based system with four channels spaced 25 KHz from each other on one carrier while TERAPOL is an FDMA.

1.3 Understanding TETRA Modes of Operation

The TETRA system enables two different modes of communication in its mobile stations, and these include TMO and DMO. Between the two modes of communication, DMO can be said to have several applications that it can run as well as having many advantages. Some of them include being operational outside the coverage of TMO, giving additional capacity when TMO networks are highly loaded, being able to work in poor signal strength areas, having the ability to get back into operations whenever the TMO infrastructure is not sufficiently operating, possessing the utility applications that can be used by organizations without the need for trunked network capacity, and the fact that communication can take place on a single carrier.

A major point to note is that the DMO is a unique system with a specific feature for TETRA as well as other specialized communication systems, and this is what makes it be different from other public and private cellular mobile networks (Stojkovic 2016).

1.3.1 TETRA Security

The level of security that has to be provided by TETRA networks to its users should be high and trustable. TETRA, TETRAPOL, and P25 among other PSN systems must always aim toward protecting the user's information, identity, and other crucial information. The security functions of these networks are often categorized into four entities, which include security mechanisms, security management aspects, standard cryptographic algorithms, and lawful interception mechanisms.

1.3.2 Evaluating the Challenge of Data Transmission and Possible Solutions on TETRA Networks

The objective of this part of the chapter is to evaluate the ongoing changes and transformations in PSNs communications. For instance, there will be a discussion on the abilities of LTE standards to replace TETRA and an assessment of the outcome, for instance, will it be better? In addition, there will be an exploration of the different transition scenarios of PSNs.

The current trends have shown that the behavior of PSNs to be changing. There is increasing need for voice-centric services to be gradually replaced by data-centric applications. As time progressed in the development, it was discovered that the utilization of applications like picture transmission, live videos, and high-speed internet among others could work in favor of public safety. The majority of specialists recognized that these data applications are capable of changing the perception of users on how PSNs enhance communication, which can ultimately improve the functionality of public safety. Whether one decides to view these systems in terms of priorities, needs, or the size of the market, it is rational to state that the public safety communication systems have been on the verge of transformation at different speeds. At one point or the other, the development has experienced stagnation in limiting the capabilities of data transmission. For instance, in most cases, we find that most public safety communication systems fail to satisfy the users' needs and offer sufficient bandwidth for the massive data applications. However, technology in the commercial domain has been evolving at a faster level, leading to the belief that commercial cellular systems are unreliable when it comes to functionalities required for the normal operational work of public safety users such as PTT features as well as device-to-device communication.

Since governments and public safety agencies have realized that changes toward the systems are inevitable, they have begun to create a solution that will integrate these systems and ensure ease in interoperability. For example, commercial LTE networks offer good support for data application but they lack the specialized support services that are important in communication. Conversely, the public safety TETRA networks have good support for specialized systems but they lack sufficient support for data applications that require a high throughput. The devised solution included the establishment of a single standard that is built to meet the user's requirements while at the same time providing specialized services that are crucial for operational purposes. Because the TETRA standard is incapable of supporting expansive broadband applications, there is a need for another solution. Specialists came to the conclusion that LTE should be used as a single global standard for the public safety telecommunications. Hence, the next step would be to ensure that LTE functions to its best potential.

As part of the solution, there have been particular improvements in LTE to build public safety (PS LTE). This type of development showcases an important turnaround not only for LTE but also an ultimate improvement in PSNs as well as the role of telecommunications. When it comes to the revolution of LTE, a big challenge has been recorded because of the pressures to meet the requirements of every user in achieving the same level of reliability and the services available and provided by the existing PSNs.

1.3.3   Comparing Public Safety Networks to the Commercial Cellular Networks

The significance of discussing this topic in this chapter is to help readers comprehend how the future public systems should be built, or rather look like. This includes looking into the characteristics they should have. In the discussion, the TETRA and LTE networks will act as the representative models, where the PSNs will be represented by TETRA and the commercial network LTE.

Over the past decades, PSNs and commercial cellular networks have displayed various needs and services depending on the technology they are run on. The three basic telecommunication technologies used in communication networks include

> ➢ Narrowband (NB) technology. This technology is designed to provide voice-centric communication and low-speed data applications. Furthermore, in the system, data rate systems are limited to probably a few tenths of a kilobit. The TETRA networks use this kind of functionality to deliver its services.
> ➢ Wideband (WB) technology. These are technological services that provide application data rates of between 384 and 500 kbps. After the launch of TETRA Release 2, the system has strived to improve the data services by establishing the TEDS. This TEDS liaises with the wideband technology to give higher data rates, the highest probably reading 500 kbps (TCCA 2018). However, the usage of this technology in the market has not been aggressively embraced because its data rates are not so high to support applications with higher bandwidths such as video conferencing and video streaming that might require up to tenths of Mbps and 10 Mbps respectively.
> ➢ Broadband (BB) technology. Now this is the type of technology for communication that is famous for supporting bandwidth-hungry applications. It is capable of supporting higher-speed data communications as compared to the wideband. The data rates of such technologies can go up to 300 Mbps in the LTE networks. As is widely known, LTE networks utilize broadband technologies for their services. The downlink data rated can record up to 100 and 300 Mbps, whereas on the uplink they range from 50 to 75 Mbps.

1.3.3.1   Services
TETRA shares particular basic technologies aspects with cellular mobile networks, but with additional critical features. Major characteristic of public safety users is that they work in groups. In addition, the communication systems used in these types of networks have a specific design to ensure the optimization of various services provided. On the other hand, commercial cellular communication systems were designed for person-to-person communication, making them ineffective in supporting public safety communications. For public safety users, they can use DMO to communicate outside the stipulated network coverage (Stojkovic 2016; Stavroulakis 2007).

1.3.3.2   Networks
Table 1.1 presents a summary of the differences between a commercial network operator and PSN models while attempting to discover the future technologies that can be implemented on the latter to enhance performance and functionality. The table was aided by the work of Ferrãos and Sallent.

Table 1.1 Commercial networks and PSN models

Issues	Commercial network operator model	Public safety network model
Goals	Maximize revenue and profit	Offer protection to life, property, and state
Capacity	Often defined by a "busy hour"	Often defined by "worst case scenario"
Coverage	Population density	It is territorial, focuses on whatever may need protection
Availability	Outages are undesirable	Outages are unacceptable because lives may be lost
Communications	One-to-one	Encompasses dynamic groups
Broadband data traffic	Internet access mainly downloads	More uploads as compared to downloads
Subscriber information	Ownership is identified as a carrier	Ownership is by an organization
Prioritization	Often has minimal differentiation based on the subscription level or the application	There is significant differentiation by role or the incident level
Authentication	This is carrier controlled, and device authentication only	It is organizational controlled with user authentication
Preferred charging method	There are subscriptions with a predefined number of minutes, for example, per minute for voice, per GB for data, per message for SMS; all these with a fixed price	There can be quarterly or annual subscriptions with limited use

These differences indicate that the future LTE will have to fill the gaps and overcome any limitation presented by the PSNs.

1.3.4 How to Overcome These Differences

1.3.4.1 Limitations of TETRA

Even though the initial design for TETRA was to offer many specific services, being reliable and ensuring security, these systems have now been considered to be outdated to survive in the modern technologically advanced world. Mainly, this is because of their insufficiency in supporting advanced broadband applications because the technology these networks use limit them from performing particular functions. A second limitation identified by the TCCA for the network system was its inability to support modern data applications (Stojkovic 2016).

In April 2002, a working group of the TCCA named the Critical Communications Broadband Group (CCBG) came up with a mission: "drive the development of one or more common standards of mobile broadband that fulfil the mobile applications needs of users who operate in a critical communication environment and will lobby for appropriate harmonized spectrum in which they deploy such services" (CCBG 2012).

Later in October the same year, TCCA produced a report stating that LTE had been selected and granted the mandate to deliver mobile broadband solutions for the users of mission and business critical mobile communications.

1.3.4.2 Need for Broadband

The majority of users were discovered to be moving from voice-centric communication toward information-centric operations as the latter offered various ways to share information. Additionally, the report indicated an introduction of broadband data application such as the internet/intranet access, video streaming, and web browsing would greatly enhance the functionality of the PSNs (Mason 2010). Besides this report, there have also been other studies that show that the deployment of broadband technology has a really great potential in generating indirect returns as well as other socioeconomic advantages (Mason 2010).

1.4 Unifying the Two Worlds of Public Safety Networks and Commercial Networks

Over the years, there have been indications that both these networks have experienced advancements and both have advantages. The unification of these two networks will surely bring major advancements in the evolution of public safety. As stated by Ferrãos and Sallent (2015a,b,c) major transformation will, of course, require a merger or migration of TETRA and TETRAPOL to the LTE networks although this will not be an task to achieve because of various special requirements to meet the needs of the public safety users. Ideally, the future PSNs should ensure they have the exact level of control, security, and high broadband as well as ensuring the modern applications are well supported (Stojkovic 2016).

1.4.1 User Requirements

Even though the LTE networks are attractive, transiting TETRA to it is quite challenging because of the various standards the networks have to meet. A research by TCCA showed that there are various requirements that are needed to successfully transit TETRA to the modern LTE networks.

This means that the future public networks must guarantee the following services:

➢ Group communications. This means that there should be communication across the group of users and other relating services such as dynamicity in groups, late entry, and group management, amongst others.
➢ PTT service. Here, communication happens over the mobile radio network and the users just need to press the talk key to activate the voice transmission path before they can start using it.
➢ Device-to-device communication. This refers to the communications between mobile devices that are independent of the network.
➢ Prioritization and pre-emption. This requirement means the ability to enable the most critical or important calls to be connected first and fast in times of congestion.
➢ Emergency calls. This basically means having prioritization above other traffic

Moreover, regardless of the technology, the PSNs need to have certain characteristics that will ensure they operate as they should. These include:

> Coverage. For example, radio coverage should be about 100% in the geographical area it is covering, Moreover, it should have the ability to operate outside the stipulated networks it is supposed to cover under device-to-device communication.
> Scalability. This encompasses an array of concepts such as different cell loads, symmetric uplink/downlink usage pattern, and how services become available at different speeds.
> Availability. The network resilience for these networks as well as service availability should be constant throughout the year, probably close to 100% with a downtime of fewer than five minutes annually.
> Interoperability. This is basically the ability of the network to integrate and communicate with other networks.

1.4.2 Public Safety Network Migration

The personal evolution of LTE is very important to become the best PSN in the modern world. This entails an array of factors but the most significant one ensures it meets the user's requirements. As stated by Ferrãos and Sallent (2015a,b,c) three techno-economic aspects are very crucial in this particular transition. They include a technology dimension, a network dimension, and a spectrum dimension. All these approaches have significant roles in defining the LTE as a modern PSN system and a future mobile broadband technology.

When it comes to the technology dimension, the focus is on 3GPP as an enabler of LTE to become the future of mobile broadband technology. From the technological dimension perspective, there should be necessary functionalities in LTE systems to ensure the provision of services that are only currently available in TETRA. Secondly, the network dimension can be explained in term of business and delivery models that are suitable for LTE to provide better PSN. On the other hand, the last dimension, which is the spectrum, can be perceived in terms of the several regulations that should be followed at the local and global level to ensure uniformity of services.

1.4.3 Deployment Models

Based on the previous discussion, it is rational to state that the evolution of PSNs will also heavily rely on the deployment criteria and the respective business models to ensure efficiency and effectiveness. Hence, different deployment models should be taken into account as well as the implementation options. Some of the proposed models identified by the specialists include:

> LTE dedicated networks. These are networks established for public safety communications. They are specifically designed to meet the requirements of the users.
> LTE commercial networks. These refer to commercial networks that can also aid in public safety communications
> Hybrid solutions. This refers to the integration of both the LTE dedicated networks and LTE commercial networks, or a combination of the two, as well as legacy PSNs such as TETRA.

1.5 The Transition from TETRA to LTE and the Current Initiatives

Over the past few years there has been wide recognition of the advantages of having broadband communication in many government agencies, and some countries have even started to prepare to embrace mobile broadband public safety communication systems. Some of the leading countries in this transition include Australia, Canada, New Zealand, the United States, South Korea, the United Kingdom, and Belgium (Stojkovic 2016). Each country has their preferred deployment model. For example, the United States is establishing an LTE dedicated network for public safety use, whereas the Belgians prefer adopting the LTE commercial network to offer a data-centric application for its users (Stojkovic 2016). Moreover, other countries thinking of embracing the Belgian model include Finland and France. Just to mention a few of the models, the United States has FirstNet, the United Kingdom has ESN, and Belgium has the government-owned operator ASTRID (Ferrãos et al. 2013).

1.5.1 Network Softwarization

The networks we are currently using in the modern world have become so complex that it has become necessary to rethink ways to increase their efficiency as well as make their management easier. One of the proposals that have been made to achieve this includes the establishment of SDN (Wetterwald et al. 2016). In this approach, there needs to be a clear separation between the data plane and control plane in the network logic. OpenFlow is the most significant feature of this system because it helps in the prioritization of communications. In addition, some of the virtual functions of SDN enable rapid deployment of PSN leasing to strong control planes thereby ensuring the entire system is robust. Basically, this is one of the systems that can be used to revolutionize PSNs.

1.5.2 LTE Technology for Public Safety Communications

LTE has been widely recognized as the first cellular communication technology that has hugely redefined the technology footprint of the mobile industry after it succeeds the Global System for Mobile Communications (GSM), Code Division Multiple Access (CDMA), and the Universal Mobile Telecommunications System (UMTS). The standard of LTE has been embraced globally making it supported both technically and economically. In addition, its ample benefits, such as low latency and high bit rates amongst others, are the reasons why it is increasingly becoming a popular network.

The success rate of LTE has been glorified in the world of public networks and it is not surprising that TCCA chose it to replace TETRA technology. Its standard for the mobile broadband has mounted pressure on this kind of network to improve. According to Stojkovic (2016), these proposed improvements would ensure the networks are offered a reasonable solution, and they would ensure excellent "functionality in group communication, device-to-device communication, push-to-talk service, and isolated and independent work of base station through the dispatch and talk through mode."

1.5.3 LTE as a Public Safety Mobile Broadband Standard

The standardization for LTE within the 3GPP began in 2004 with the launch of Release 8 and some minor changes in Release 9 launched in 2008. More specifically, the first public safety in LTE was the beginning of Release 11, which was an active system from 2010 to 2013. At this juncture, the use of LTE as a PSN was not so much popular until after the launch of Release 12 and 13 developed from 2011 to 2016. Figure 1.1 represents the 3GPP Releases timeline and what the future holds for LTE up to 2020.

1.5.4 Security Enhancements for Public Safety LTE Features

In any kind of technology, security plays a really important role. In the case of PSNs, the nature of security is on another level because the original purpose of these networks is to ensure a safe and stable environment while observing the law and, most significantly, protecting human life. This means that security in these systems has to be of a high priority. Past studies have often categorized security features to fall under two categories: individual security procedures and common security procedures.

The common security procedures include an array of features such as having a network domain security, security of functional interfaces, and security of the PC2. On the other hand, individual security procedures include developing security direct discovery, security for one-to-many ProSe direct communication, security for the evolved packet core (EPC)-level discovery of ProSe-enabled UEs, and security for EPC supported WLAN direct discovery and communication (Stojkovic 2016).

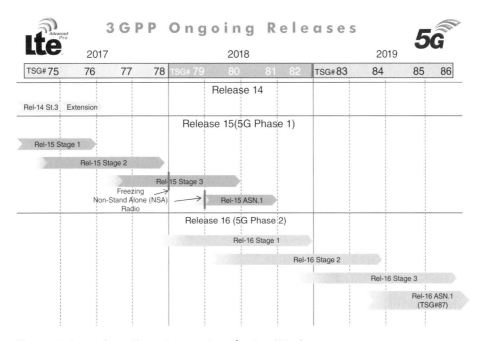

Figure 1.1 Source: https://www.3gpp.org/specifications/67-releases

When it comes to the security of the PSNs, some of the proposed establishments include an MCPTT application plane security by engaging in MCPTT user authentication, Session Initiation Protocol (SIP) registration and authentication, and MCPTT service authorization (Stojkovic 2016).

1.6 Conclusion

It is clear that the evolution of PSNs from TETRA, TETRAPOL, and P25 toward LTE based systems have shown effectiveness and efficiency. Even though the former systems have various advantages in terms of communication, the evolving LTE systems have proven to be far better, especially in terms of broader bandwidths that can support various applications, hence making the technology a promising feature for the PSNs. Currently, many public networks are shifting toward the existing technologies thereby equipping their users with the capabilities that are similar to the commercial data networks. Some of the approaches discussed in this chapter that will take the PSNs to a whole new level include softwarization and virtualization as defined in SDN. These software networks assist in defining the clear separation between the data and control plane, meaning that there will be a quicker convergence in the newly installed networks, a centralization of policy decisions, and flexibility in reconfiguring network functions whenever there is overload or hardware damage. Furthermore, the study shows that the new LTE functionalities that are in the process of being designed for public safety communications to have the potential of replacing TETRA by providing similar functionalities, are at an advanced stage. These new implementations will mean that the LTE architecture will have to get more network entities, especially application servers, because they will assume this role. With many small technologies added to LTE, the future LTE systems services will match or surpass those of TETRA and ensure they meet the public safety user requirements for the group and device-to-device communication by GCSE-LTE, and ProSe functionalities. Moreover, the PTT service can be utilized to assist in MCPTT in the new technology.

References

CCBG, (2012). "Mission/Vision Statement," 23 April 2012. http://www.tandcca.com/ Library/Documents/CCBGMissionv1_0.pdf. (Accessed 29 March 2016).

Chavez, K.M.G., Goratti, L., Rasheed, T. et al. (2015). The evolutionary role of communication technologies in public safety networks. In: . DiVA http://www.diva-portal.org/smash/record.jsf?pid=diva2%3A930311&dswid=7213.

Doumi, T., Dolan, M.F., Tatesh, S. et al. (2013). LTE for public safety networks. *IEEE Communications Magazine* 51 (2): 106–112.

Favraud, R. and Nikaein, N. (2015). Wireless mesh backhauling for LTE/LTE-A networks. In: *Military Communications Conference, MILCOM 2015–2015 IEEE*, 695–700. IEEE.

Ferrús, R. and Sallent, O. (2015a). *Mobile Broadband Communications for Public Safety: The Road Ahead Through LTE Technology*. Wiley.

Ferrús, R. and Sallent, O. (2015b). Future mobile broadband PPDR communications systems. In: *Mobile Broadband Communications for Public Safety: The Road Ahead Through LTE Technology*, 81–124.

Ferrús, R. and Sallent, O. (2015c). LTE networks for PPDR communications. In: *Mobile Broadband Communications for Public Safety: The Road Ahead Through LTE Technology*, 193–256.

Ferrús, R., Pisz, R., Sallent, O., and Baldini, G. (2013). Public safety mobile broadband: a techno-economic perspective. *IEEE Vehicular Technology Magazine* 8 (2): 28–36.

First Responder Network Authority Reston, VA, USA First Responder Network Authority (2012). Why Firstnet, https://www.firstnet.gov/network/10-ways

Mason, A. (2010). Public safety mobile broadband and spectrum needs. Final report. Newcastle upon Tyne TCCA

NOKIA. LTE networks for public safety services. White Paper 2014

Stavroulakis, P. (2007). *Terrestrial Trunked Radio-TETRA: A Global Security Tool*. Springer Science & Business Media.

Stojkovic, M. (2016). *Public safety networks towards mission critical mobile broadband networks* (Master's thesis, NTNU).

TCCA. (2018). "TETRA Release 2," TCCA, Newcastle upon Tyne TCCA. Available online: http://www.tandcca.com/about/page/12029. (Accessed 20 June 2018)

Wetterwald, M., Saucez, D., Nguyen, X. N., & Turletti, T. (2016). *SDN for Public Safety Networks* (Doctoral dissertation, Inria Sophia Antipolis).

2

Public Safety Networks Evolution Toward Broadband and Interoperability

2.1 Introduction

Founded in 2003, the Department of Homeland Security (DHS) in the United States has made it a priority to enhance communication with the Nation's emergency responders, seeking real-time information sharing among responders during all threats and hazards to the safety of Americans. The National Emergency Communication plan focuses on sharing information with the public and private sector entities, government entities, and non-government entities. The various stakeholders' engagement with American safety made the DHS' first plan a success.

The safety of citizens depends on the ability of first responders to communicate and formulate an applicable approach to resolving a disaster. Development of appropriate communication platforms to provide a system that can interlink between several agencies as occurs in the commercial platform is essential. Problems in public safety communication systems are notable for hindering different geographies and agencies in first response to communicate and coordinate.

This chapter looks at how public safety networks have made efforts in order to advance from narrowband to broadband systems for a better services and interoperability. These public safety networks are a collaborative effort of the government, technological industry, and a number of other interested parties like telecommunications companies. The collaboration has seen to the improvement of public services that might be in line with making sure that there is unified global advancement toward using broadband systems where many entities can communicate and share information. More light will be shed on the broadband system architecture, its benefits as well as how public safety and security can employ its services in ensuring that people are accorded with standard service as part of public safety.

2.1.1 Communication Technology

Looking back to the nineteenth century, it is evident that the means of human communication has undergone significant transformation. Unlike back then, when communication of information was reliant on pen and paper, contemporary improvements in communication technology do the same just by the click of a button. Innovative technologies in communication have made it possible to attain connection between people and businesses by changing the manner in which people interact on a daily basis. Communication technology places emphasis on the role of integrated communications as

Public Safety Networks from LTE to 5G, First Edition. Abdulrahman Yarali.
© 2020 John Wiley & Sons Ltd. Published 2020 by John Wiley & Sons Ltd.

well as the incorporation of telecommunications, storage, audio-visual systems, and enterprise software, which makes it possible for users to gain access to storage, transmission, and management of information. It is due to communication technology that integrating computer network systems with the telephone network systems has been promoted. This is achieved through the use of a single amalgamated system of signal distribution, cabling, and management.

Communication technology has the ability to impact on society and business operations through enabling efficient exchange of ideas and information. Communication technology encompasses e-mail, the internet, telephone, multimedia as well as other video based and sound based means of communication. Specialists in the field of communication technology are responsible for structuring and maintaining technical systems of communication with regards to the needs of industry, business, or market. Communication technology has the potential to considerably impact on the execution of an assignment in order to ensure swift and steadfast communication between team members.

A perfect example would be e-mail, which has become a standard means of official communication for the majority of organizations. Using e-mail to communicate is effective when it comes to dealing with numerous customers, stakeholders, and partners. All these parties involved can be accessed through sending one e-mail. Aside from e-mail, text messaging is also another example of communication technology that has proved effective when it comes to communicating information in urgent situations. Instant messages can be sent through various applications and websites that significantly contribute to successful business negotiations. Like e-mail, text messaging has the ability to engage two or more stakeholders in different geographical locations. It would be unfair not to mention social networking sites like Twitter, MySpace, and Facebook. These sites have amounted to being the preferred means of communicating information to particular clients or targeted public. Most corporations have set up official pages on social networking sites for the purpose of associating with their clients.

2.1.2 Wireless Communication Systems

A communication system, Figure 2.1, comprises of a number of components: an input transducer, a transmitter, a channel that is affected by distortion and noise, a receiver, and an output transducer. The difference between a wireline communication system and a wireless communication system comes in at the channel component. For a wireless channel, there is less protection from external interceptions and multi-path propagation properties. There are so many examples of wireless communication systems inclusive of cellular telephone systems, codeless telephones, hand-held walkie-talkies, paging systems, and garage car openers.

Wireless communication systems are punctuated with significant trends. These trends include the convergence of wireless and internet broadband communications and prompt growth as a result of large-scale circuit integration, digital signal processing, digital switching techniques, and digital and RF circuit fabrication. 1G wireless communication systems made their first appearance in the late 1970s with low system capacities and backbone structures grounded in analog techniques. In the mid-1990s, 2G wireless communication systems were deployed characterized by digital modulation and voice-centric services. After that came the 2.5G wireless communication systems

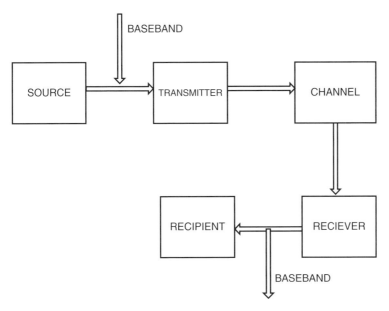

Figure 2.1 Components of a communication system.

followed by the 3G wireless communication systems. The latter was characterized by an increased transmission rate as well as the backing of multimedia services.

When focusing on broadband, the focus shifts to wireless local area networks and personal area networks. These two networks have the ability to give broadband telecommunications access in a local exchange. This provision is prompted by the demand to access the broadband internet from homes as well as businesses as a result of the steady but fast growth of the internet.

2.1.3 Government Involvement

In order to be able to compete in the modern economy, most communities have realized that they are dependent on internet access. Due to this, access to broadband has amounted to a significant requirement for the life quality of a community. Broadband helps to provide people with access to telemedicine, online education and distance learning programs, and entertainment content. In today's world, broadband has grown to become so integral that at the moment most people hold it in high esteem as a basic infrastructural requirement together with things like water systems, roads, and energy grids.

Even though the provision of broadband is not considered a service that should be offered by the local government in historical or statutory terms, local government officials have a say. Local officials play an active role in determining whether their communities are able to gain access to high-speed internet infrastructure. In the past few years, a significant number of local governments have been looking at the prospects of investing public resources in order to attain an enhancement of broadband infrastructure. These improvements come in the range of cell towers or fiber optic lines. In this way, the local government is able to entice internet service providers as well as improve efforts at economic development in the local regions.

Local governments have a significant role to play in community development through efforts at bettering the economy. This role is presented as the responsibility to procure broadband infrastructure for local communities. Having broadband services that are accessible, affordable, and redundant might prove abstract in most local communities in developing countries. However, overcoming such a challenge should not be considered as unattainable. Such communities have the prospect of building upon successful past efforts in bringing to the table non-profit, state, industry, and federal partners. In this way, local governments have the ability to successfully bridge the evident digital gap and, in turn, help communities obtain a standard broadband infrastructure that is credited for economic development in other areas.

2.2   Evolution to Broadband Systems

The world as we see it now is in the middle of stimulating growth and groundbreaking use of mobile broadband. This serves as a witness to the vast spread of mobile broadband networks, development of mobile services and the appearance of innovative mobile device classifications. All of these monumental developments are taking place at an extraordinary pace. Looking back at the early 2000s, ATM, acting as the mass technology coupled with the requirement to make accessible triple play residential services, resulted in having operators defining an innovative aggregation network that was grounded in ethernet technology (Costello 2012). This kind of architecture together with its service model acts as the base for the contemporary broadband architectures.

The history of broadband wireless systems can be linked back to the need to find an inexpensive option to the present traditional wireline access technologies. Motivated by the deregulation of the telecommunications industry as well as the fast growth in the use of the internet, a number of competitive carriers have been pushed to find a wireless solution that has the ability to bypass mandatory internet service providers. In the last 10 years, numerous wireless access systems have been developed as a motivation from the disruptive potential of wireless. The majority of this development was handled by start-up corporations. These wireless access systems had different characteristics from their protocols, the applications that they supported, performance capabilities, the frequency spectrum put to use and many other varied parameters. Some of the systems underwent commercial deployment but were decommissioned much later on. Effective deployments have so far been restricted to a number of niche markets and applications. Clearly, broadband wireless in the past has been characterized by a checkered record. This record can be attributed to the fragmentation of the industry as a result of the lack of a standard.

Unceasing growth and development of the internet are in need of evolvement of the general network architecture (Forum 2014, p. 16). The architecture of broadband networks is significantly affected by the introduction of the IPv6 backing. This reliability manifests not only in terms of the move toward IPv6 but also in terms of the co-existence of IPv4 and IPv6 in service provider networks. Taking such an environment, broadband service providers possess exclusive access to clients. The broadband service providers are able to influence and enhance the existing network architectures to be able to facilitate cloud as well as virtualized services. This ability is regarded as an essential objective for service providers that view cloud computing as an innovative fountain for revenue. Cloud computing will definitely play an integral role in engineering innovative business considering the fact that it makes it possible for broadband service providers to enter the IT services market with the help of their current assets. Cloud computing is bound to continue having a significant effect on broadband networks in the future. Cloud services that are made available by broadband networks can be classified using architecture as an innovative set of services under the multi-service group (Sidak 2016, p. 78). This multi-service umbrella has components, architecture, and an access model that can be influenced to make available cloud services.

Looking at companies that have and are still playing a significant role in the evolution toward the broadband systems, Cable AML features at the top. Cable AML is regarded as a front-runner in the growth and implementation of broadband wireless access systems. In fact, the BWA-2000 access platform made possible by Cable AML is one of the four platforms that have been developed over the past two years. The company offers a complete broadband wireless access product that enables matching between a platform and every client's needs. In addition, the use of average access platforms has given room for the whole incorporation of a small current cable plant into the prevailing wireless system. The company; Cable AML, has enabled the delivery of a two-way repeater. This two-way repeater has the ability to augment the coverage area of a broadband wireless system that is operating at 28 GHz. To add to that, Cable AML released the BWA-1028 and BWA-4028S broadband wireless access systems. These two compact fixed-wireless systems are ideally matched to enable a "last mile" two-way data service to clients that are in need of high-speed connectivity.

2.2.1 Determining Factors

- Performance levels. In order for the evolution of broadband systems to be realized and be effective, there is a need to measure the performance of the present systems. This task has been taken up as a responsibility by various stakeholders like governments. For instance, the US government has a National Broadband Plan that has as part of its activities the Measuring Broadband America program. The program makes use of the Federal Communications Commission speed test application for iPhone and android devices to measure the performance of smartphone mobile broadband services from volunteers' phones. The performance data that is being gathered concerning broadband is inclusive of download and upload speeds, packet loss and latency, and wireless performance features of the broadband link, and the types of handsets and forms of operating systems tested. The application has the ability to measure mobile broadband performance in four classifications: upload speed, packet loss, download speed, and latency. In order to make the analytical experience more efficient, a number of additional passive metrics are also recorded like the model of the device, manufacturer of the device and signal strength of the connection. In the instance that a broadband system gives signs of not working efficiently then action has to be taken by the developers. The evaluation stage has always been important for any project. Evaluation gives room for project perpetrators to make amends where needed and increase efficiency. In this way, the performance levels of existing systems determine the direction in which the evolution of broadband systems is going to take.

- Technological advancements (architecture). Since the introduction of broadband, the technological world has witnessed the development of innovative broadband technology like 4G mobile broadband (Khan 2009, p. 22). This new broadband technology makes it possible for clients to be available online in any location. There is also the example of cable broadband that has enabled the enhancement of connection speeds of up to 300 Mbps. Concerning the architecture of broadband services, there have been stimulating trends that have been developed in the past years. Part of these trends is the rapid waning of asymmetrical digital subscriber line (ADSL) services in small business sectors and consumer sectors alike. ADSL services are being replaced with fiber based services or cable modem services. The new services that have been prompted by the enhancement in system architecture have experienced a commendable uptake by consumers. When broadband was first put in the market, it was considered expensive and only a selected few could gain access to it. However, at the present time, close to everyone has the ability to gain access to broadband services. This increase in availability is attributed to the existing competition amongst internet service providers. Competition meant that every internet service provider wanted to offer an improved version of broadband service. This prompted the architectural advancements that have in turn led to the evolution of the broadband system (Kumar et al. 2010, p. 67). We have witnessed the emergence of 2G, 2.5G, 3G, 4G, and 5G over the last several decades. These technologies have the ability to provide reliable and swift mobile internet at commendable speeds. If the development pace of technological advancements had been slower or stunted then the broadband service would never have evolved to what it is at the moment.

- Governments. Broadband is considered an essential motivator of economic growth of various countries. To add to that, it is also becoming the principal mechanism for gaining access to information. Information is considered a public asset that is important for all types of economic deeds and efficient governance. Broadband networks have grown to be used for the delivery of public services like healthcare, electronic land registration, financial services, and electronic voting. In the past, governments have always played a motivating role in making sure that there exists the right setting for the provision of an information and communication technology (ICT) infrastructure as well as the enhancement of the domestic ICT sector. This shows just how much of a determinant governments are in the evolution of broadband systems. Different governments globally have taken up varied stands on the choice to put in place national policies on broadband. However, in general, countries that have taken it upon themselves to develop comprehensive national policies tend to experience more success in the development of broadband diffusion. In essence, governments, local and national alike, have the responsibility to make sure that there is equitable access for everyone and for making markets function more efficiently when it comes to broadband services. The input of governments does a lot to determine the direction that the evolution of broadband systems takes.

2.2.2 Evolution Process

Figure 2.2 shows the evolutionary path of cellular systems from 2G to 5G. The demands on quality, services, and applications by consumers has paved the road to the evolution of networks, devices, and technologies.

- General packet radio service. GPRS is a packet concerned with mobile data provided on the 2G and 3G cellular communication system's global system for mobile communications (GSM). GPRS delivers moderate-speed data allocation by means of unexploited time division multiple access (TDMA) channels. GPRS was introduced in 2000 as a packet-switched data service set into the channel-switched cellular radio network GSM. GPRS spreads the grasp of the static internet by linking mobile terminals globally. Consistent with a study on accounts of GPRS growth, it was the first system offering international mobile internet access.
- Enhanced data rates for GSM evolution. EDGE is digital mobile phone machinery that permits enhanced data conduction rates as a backward-compatible allowance of GSM. EDGE was installed on GSM networks back in 2003; at first by Cingular (now AT&T) in the United States. Using the institution of refined approaches to coding and conducting data, EDGE conveys an advanced bit rates per radio channel, ensuing in a threefold upsurge in aptitude and performance likened to a regular GSM/GPRS association. EDGE/EGPRS (enhanced general packet radio systems) is applied as a bolt-on improvement for 2.5G GSM networks. This makes it simple for current GSM

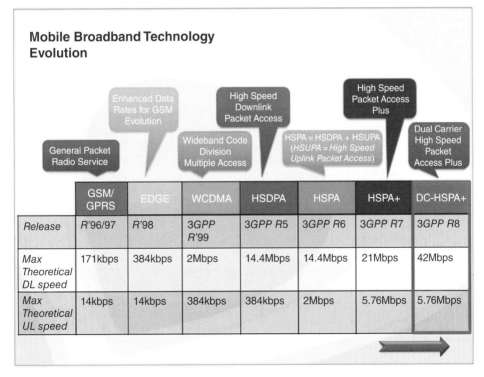

Mobile Broadband Technology Evolution

	GSM/ GPRS	EDGE	WCDMA	HSDPA	HSPA	HSPA+	DC-HSPA+
Release	R'96/97	R'98	3GPP R'99	3GPP R5	3GPP R6	3GPP R7	3GPP R8
Max Theoretical DL speed	171kbps	384kbps	2Mbps	14.4Mbps	14.4Mbps	21Mbps	42Mbps
Max Theoretical UL speed	14kbps	14kbps	384kbps	384kbps	2Mbps	5.76Mbps	5.76Mbps

Figure 2.2 Evolution of mobile broadband technology. Source: Accessed from http://www.techcrates .com/wp-content/uploads/2018/01/evolution-of-mobile-broadband-technology.jpg.

movers to elevate to it. EDGE is a superset of GPRS in addition to having the ability to work on any network with GPRS positioned on it, on condition that the carrier carries out the required advancement.

- Wideband code division multiple access. WCDMA is a 3G standard that services the direct-sequence code division multiple access (DS-CDMA) channel access technique and the frequency-division duplexing (FDD) technique to make available high-speed and high-capacity provision. WCDMA is the most frequently used alternative of the Universal Mobile Telecommunications System (UMTS). It was established by Japan's NTT DoCoMo and shaped the foundation of its Freedom of Multimedia Access (FOMA) 3G network.

- High-speed downlink packet access. HSDPA is an improved 3G mobile communications procedure in the high-speed packet access (HSPA) category, also named 3.5G, 3G+, or Turbo 3G, which agrees with networks grounded on UMTS to possess advanced data speeds and aptitude. HSDPA is founded on a common channel broadcast and its main characteristics are a mutual channel and multi-code broadcast, higher-order variation, little transmission time interval (TTI), swift link adaptation and scheduling along with swift hybrid automatic repeat request (HARQ).

- High-speed packet access. HSPA is an incorporation of two mobile procedures: HSDPA and HSUPA (high-speed uplink packet access), which spreads out and advances the results of 3G mobile telecommunication networks by means of the WCDMA procedures. The original HSPA stipulations sustained augmented peak data rates of up to 14 Mbps in downlink and 5.76 Mbps in uplink. In addition, they also condensed inactivity and delivered up to five times additional system capacity in downlink as well as up to twice as much system aptitude in uplink, similar to the original WCDMA procedure.

- High speed packet access plus. HSPA+ is a practical standard for wireless, broadband telecommunication. It is the subsequent stage of HSPA that was made known in 3GPP Release 7 and was extra upgraded in later 3GPP releases. HSPA+ has the ability to attain data rates of up to 42.2 Mbps. HSPA+ is an advancement of HSPA that improves the 3G network and makes available a technique for telecom machinists to drift in the direction of 4G speeds that are additionally equivalent to the initial accessible speeds of newer Long-Term Evolution (LTE) networks short of installing a new radio (NR) crossing point (Taylor 2014).

- Dual carrier high-speed packet access plus. Further releases of the HSPA+ standard have presented dual carrier operation, which is the concurrent use of two 5 MHz carriers. The DC-HSPA+ is a fragment of the 3GPP Release 8 description. It is the regular development of HSPA by way of carrier combination in downlink. UMTS certificates are every so often issued as 5, 10, or 20 MHz combined spectrum distributions. The straightforward impression of the multi-carrier characteristic is to attain improved resource consumption and spectrum competence by way of combined resource distribution and load harmonizing across the downlink carriers.

2.2.3 Broadband System Architecture

The mobile wireless communications industry is one of the largest, fastest growing industries in the history of US broadband wireless and is a significant growth marketplace for the telecom industry to deliver a variety of applications and services to both mobile and fixed users. Figure 2.3 shows a general schematic diagram of a broadband network architecture.

- Fixed broadband. Fixed broadband denotes that equipment where the end user is obligated to remain at an unchanged position to make use of the broadband service. The access network is connected to the precise physical locality. Fixed broadband comes in a number of forms: wireless fixed broadband, wireline fixed broadband, and satellite fixed broadband. The first one, wireless fixed broadband, has worldwide interoperability for microwave access as the most common technology. This popularity is contributed to by affordability when put up against other options like fiber access. The second one, wireline fixed broadband, comes in different types like the cable fixed wireline broadband where the broadband service is acquired by means of cable access by improvement of a traditional cable television network, DSL fixed wireline broadband, which is the most shared access network technology, and fiber fixed wireline broadband, which may come as fiber to the premise, fiber to the home or fiber to the building. The last one, satellite fixed broadband, is in general put to use in countryside areas where no alternative access network preferences exist. However, satellite connection possesses a much advanced potential when equated with other fixed broadband access technologies.
- Mobile broadband. Mobile broadband denotes those technologies where the end consumer has the ability to use the broadband service despite the fact that they may be on the move as well as from any physical locality. At the moment, as from 2017, the most common mobile broadband technologies are third generation (3G) and fourth generation (4G) (Ekström et al. 2006, p. 42). However, this does not mean that fifth generation (5G) should be forgotten. These technologies are a means of offering diverse service speeds to the consumers and service provider access. Their backbone infrastructure is intended in a wholly dissimilar manner. Mobile broadband is the promotion term for wireless internet access conveyed by means of cellular towers to computers and further digital devices by means of transportable modems. By the conclusion of 2012, there were approximately 1.5 billion mobile broadband payments increasing to a 50% year-on-year frequency. By the end of 2018, mobile subscriptions in the world reached 5.1 billion and they are projected to increase by 700 million by 2025 (GSMA, 2019).

Incessant development of the internet requires the general network architecture to develop: broadband network architectures are intensely affected by the institution of IPv6 backing, not only in light of movement in the direction of IPv6, but also in terms of co-existence of IPv4 and IPv6 in service provider networks (Ergen 2009, p. 48). The Broadband Forum began focusing on IPv6 migration in 2008, outlining the development of the broadband network architecture from an IPv4 only network to a dual-stack IPv4/IPv6 network. As the IPv6 procedure is not backward-attuned with IPv4, a lot of effort has been put into diversifying IPv4 to IPv6 migration types of machinery in order to permit a swift migration from IPv4 to IPv6.

The Dynamic Host Control Protocol (DHCP) modus operandi was presented as a substitute for PPP (point-to-point protocol) for IP address configuration and service delivery. The use of DHCP in broadband networks began to be prevalent, and DHCP centered sessions (demarcated in Broadband Forum vocabulary as IP sessions) turned out to be a substitute for PPPoE (point-to-point protocol over ethernet) sessions for domestic

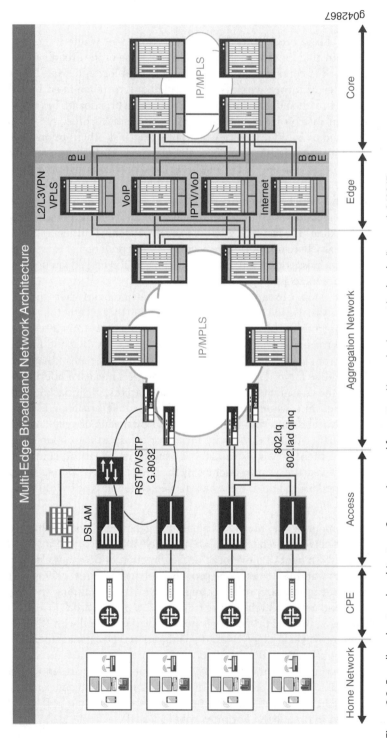

Figure 2.3 Broadband network architecture. Source: Accessed from https://www.juniper.net/techpubs/images/g042867.png.

services in situations where "always on" connectivity and modest IP centered architecture were significant necessities. Service suppliers have requested the interworking and, in some circumstances, coming together between the Broadband Forum and 3GPP networks. The Broadband Forum reacted to these requests by focusing on making parallel the telecom industry by describing the interworking requirements between 3GPP Evolved Packet Core architecture and the Broadband Forum architectures. The fundamentals of the multi-service broadband architecture are given in three essential official papers: TR-144 "Broadband Multi-Service Architecture & Framework Requirements" (DSL Forum technical Report), WT-178 "Multi-service Broadband Network Architecture and Nodal Requirements" and TR-145 "Multi-service Broadband Network Functional Modules and Architecture" (Broadband forum, Technical Report).

The important incentives that drove the description of the multi-service broadband network architecture can be described briefly as follows: interpretation of network architecture, continuous connectivity, multi-service backing for domestic, commercial, sales, wholesale, permanent, moveable, emerging cloud/virtualized services over a mutual network architecture, operational improvements, service individuality, multi-point services like those for business connectivity, multi-edge backing, improved accessibility with resiliency, and upgraded scalability.

2.2.4 Advantages of Broadband Systems

- Connection speeds are commendable; up to 100 times faster when compared to dial-up connections. It is possible to download software and picture files in just a matter of seconds instead of waiting for hours. It is due to such speeds that online gaming has been made a reality and an efficient one at that. The speed of a broadband connection is the most important icing on the cake (Casteele 2010). The data broadcast rapidity of the connection is more than 200 kbps whereas dial-up is 56 kbps. This is evident in the majority of developed countries, as many connection speeds are "provisioned." Seeing as the connection is swift, the download frequency is also swift. Consequently, transferring, music, files, emails, is much quicker with a broadband connection.
- Broadband systems do not significantly impact on the operation of phone lines. In the case of DSL internet access, it is possible to use the identical phone line to carry out both data transmission as well as voice calls or faxes. In the case of cable internet access, it is possible to be connected to the internet using the cable network. Whatever the case, the phone line in use will not in any way be occupied while the user is connected to the internet. Broadband internet access permits the internet to be linked even though the phone line is being used. This is because both the voice line and the data line are distinct in the broadband piecing together.
- The system has the advantage of having the "always on" characteristic as the internet connection is forever on. Broadband linking is "always on" with no wasted time waiting for dial-up connectivity in the case that the user makes use of the same route to the internet. This culminates in increased productivity without the need to sit around and wait for a connection to be made. The broadband connection is an "always on" connection. The broadband connection is on the moment the computer is opened. This saves a significant amount of time as compared to dial-up where the consumer has to open a phone line, key in a password, link, and then surf.

- With the broadband system, under no circumstance will the user ever dial an access number and face the possibility of getting a busy signal.
- With broadband internet, the user is assured of unrestricted access with no additional charges that are based on the duration of the connection made.
- Aside from offering high-speed internet access, the broadband system also has the ability to offer affordable phone services through voice over internet protocol (VoIP) technology.
- Considering the fact that broadband services are offered at a fixed price, it is possible for the end consumer to have an effective budget.
- Through the use of broadband internet access, consumers are now exposed to media-rich websites. This means that the majority of these websites, in addition to having a few graphics and texts, also have streaming video and sound as well as interactive elements. Such features can only be accessed through the use of broadband-based systems.
- Cheap telecoms – even though it might be relatively easy to travel around the world, the cost of making phone calls to friends in distant places can be restrictive. In answer to this, there is the VoIP, which is an innovative technology that makes it possible for the user to execute phone calls with the use of broadband internet connection in place of the analog phone line. By employing the services of the VoIP coupled with the use of modest earphones and microphones, it is now possible to make PC to PC calls from anywhere around the world for free. Even with the choice of making phone calls to a landline, the savings margin is still significant compared to the traditional method.
- Remote working – urban areas are characterized by many people who commute to and from work on a daily basis. This can be tiring and time-consuming. Due to this, more corporations are now moving toward taking up the flexibility that comes with permitting some of their staff to carry out their official duties from the confines of their homes. As a result of the available high-speed connectivity, it is possible to obtain office facilities while at home through telephone networks and internal computers. This has become even more pronounced and enticing owing to the present increase in fuel costs and corporations becoming increasingly aware of their carbon footprint. To add to that, employees that have care commitments like young children have the opportunity to restructure their working days thanks to faster broadband internet access. The advantage is even more pronounced for people who have the responsibility of running their own businesses as they are able to do it from wherever they are. This makes it possible to cut down on overheads through evasion of renting out office premises and cutting down on commuting due to work. Broadband internet access has also made it possible to use websites as a gateway to businesses. In this way, companies are able to carry on with their businesses in addition to becoming even more competitive in their respective industry.
- When using broadband internet access for things other than text communication, the consumer does not need to constantly check their e-mail application. As long as the computer is on, the e-mail application has the ability to automatically update from the server, which is achieved through checking for any new emails.
- The broadband system makes it possible for multiple users to carry out work at the same time. By contrast, dial-up connections, having multiple computers connected and being used on the same bandwidth, will result in very low speeds.

- The broadband system is characterized by a digital connection that is highly dependable. The connection experiences fewer or no breaks at all since the data transmission is efficient. In the case of dial-up modems, there are instances where data overload is experienced. In a broadband internet connection, there are no reflex cut-offs or time-outs of the connection.

When looking at all these benefits, it is evident that most if not all are grounded in ensuring public safety and security. This is achieved through the close collaboration of telecommunication companies, governments, and other corporations that make use of broadband internet access. In 1997, US Congress approved the Balanced Budget Act, assigning 24 MHz of 700 MHz spectrum to public safety. In 2007 the FCC christened the Public Safety Spectrum Trust (PSST) as the body to serve as Public Safety Broadband Licensee (Program 2011, p. 7). When talking about the advantages of broadband internet access, it is inevitable that some light is shed on how public safety and security benefits. In order to do so, this chapter will focus on the use of LTE standard and fifth generation (5G) standards for public safety and security.

- LTE based PSS system. LTE is a standard for high-speed wireless communication meant for moveable devices as well as data terminals, grounded on the GSM/EDGE and UMTS/HSPA technologies (David Astély et al. 2009, p. 48). The standard was technologically advanced by the 3GPP (3rd Generation Partnership Project) as well as being indicated in its Release 8 document sequence, with slight improvements defined in Release 9. LTE has the backing of the PSST and was suggested in the National Public Safety Telecommunication Council (NPSTC) Broadband Task Force Report.
- 5G based PSS system. The most current 3GPP standard that consists of any network exhausting the New Radio (NR) software. 5G New Radio can take account of lower frequencies, from 600 MHz to 6 GHz. Nonetheless, the speeds in these lower frequencies are modestly advanced compared with innovative 4G systems, projected at being 15–50% quicker.

Picture public safety and security systems that employ the use of either of these protocols. Such systems have the advantage of being closer to broadband internet access. This is an addition to increased efficiency in most public services that would have otherwise have been sluggish if carried out without the help of the protocols. Using the example of the healthcare industry, 5G makes it conceivable to bring reliable, steadfast user experiences to advance medical care. This is realized when carrying out things like imaging, diagnostics and data analytics, and treatment. To add to that, 5G has had a profound effect when it comes to healthcare quality, its cost, and access. On the other hand, LTE is a comparatively fresh standard for wireless communications, embraced by commercial and public safety consumers. LTE is the standard that has been accepted for the Public Safety D-Block 700 MHz frequency spectrum. An LTE network would function in a parallel manner to the commercial carriers' 4G cellular communications network. There are a number of benefits to be realized from public safety network partnerships (Williams et al. 2009, p. 18):

- Countrywide network.
- Economies of scale for user devices grounded on commercial capacity.
- Co-location of public safety apparatus on base sites and towers.
- Rapid network build-out motivated by the commercial marketplace.

2.3 Interoperability

Active shooters in small and large communities, natural disasters, and terrorist attempts require national responses. Interoperable communications have helped law enforcement, emergency medical services, and fire teams to successful protect property, the environment, and stabilize communities following responded emergencies. The success of the National Emergency Communications Plan has been a factor of various municipal collaborations including the private sector actors, and not the government alone.

An occurrence of disaster situations threatening public safety requires close collaboration between the agencies concerned in rescue (Federal Communications Commission 2010). At these times, information flowing from different stakeholder platforms require integration to enable a seamless interaction of otherwise independent agency approaches. The situations require collaboration between agents across departments and jurisdictions. Interoperability becomes difficult without the presence of necessary technology to facilitate telecommunication based on previously agreed upon mechanisms. The inability to communicate smoothly among first responders to emergencies creates delays and confusion, impeding timely rescue and raising the question of whether they are able to respond to situations.

Interoperability plays a significant role in ensuring collaboration among government and other first responder agencies with their members to prevent extensive disasters. The interoperability occurs through efficiency in communication channels to prevent and protect the public from disasters in case of occurrence. However, the data sharing among public safety networks existing is questionable in the capacity to increase inter-agency cooperation.

2.3.1 Developing an Interoperability Public Safety System

The use of cutting-edge technologies to enhance communication will enable dependability and resilience to increase operability during a disaster. Such a system requires development in clear and concise terms to address the existing problems in the field based on existing documentation. The system development needs approval from the stakeholders to allow acceptance based on the ability to communicate roles and responsibilities among the first responders. It should be practically applicable to jurisdictional circumstances.

Planning for development of an interoperability public safety system begins with the implementers agreeing to need areas that require preference. Collaboration is enhanced through the inclusion of a variety of stakeholders from several jurisdictions and departments to enable increased acceptance upon completion. Discussions should incorporate first line and junior officers to ensure acceptance in rolling out and diversity of views appropriate in each field. A sponsor to guide the decision process is essential in navigating the efforts through hurdles that could impede efficiency. Assessing needs will enable the developers to formulate a communication strategy based on the users and appropriate times to use the communication tools. Defining the level of interoperability required at each level increases understanding of what is required at each stage. Identifying the solutions requires an evaluation of the existing technologies and their level of interoperability to enable formulation of solutions to improve on the current situation. Seeking solutions based on examination of existing communication resources allows gap identification. It leads to a formulation of appropriate ways of reducing the short-comings to effective communication. Developing an appropriate technology plan will raise the level of interoperability.

Incorporating the stakeholders ensures allocation of roles and responsibilities based on capabilities. Also, regular meetings to discuss progress on plans will enable issues that arise as the communication technology is developed to be addressed. Regular reviews will enhance dedication to the initial plans of interoperability with a view to the future. The discussed plans will be implemented through the development of new procedures of approach to ensure public safety in disasters. Putting the plan into practice requires adequate training of stakeholders to enable them to use the equipment in response to emergencies. Interoperability needs a follow-up on the state of equipment on a regular basis.

2.3.2 Platform and Technology

Interoperability in emergency response occurs across various levels such as operational, legal, and political levels to achieve close collaboration (Adam et al. 2008). The existence of highly fragmented systems of communication among public safety agencies amounts to widespread and independent acting whenever a disaster threatens the lives of the public. Each administrative network of first responders employs different technology in communication, making it difficult to coordinate with other service providers during an emergency (Bedford and Faust 2010). Interoperability is beneficial in such situations as it calls for the development of a single cohesive channel of communication across all geographical locations. The presence of a single cohesive network communication path allows information sharing with rising reliability. Interoperability will reduce the impact of distorted communication through raising the dependability of information shared through the available infrastructure.

Lack of interoperability among the public safety networks is problematic. It under-mines communication whenever different agencies attempt to cooperate or share infor-mation about what is required. Increased use of multiple incompatible channels of infor-mation sharing creates delays in response capacity when the integration of the networks is impeded by structural barriers. Incompatible communication paths reduce interoper-ability by creating problems and in the process reduce the dependability of public safety networks. Existing public safety communication devices are limited in communication only with the system from the offering agency and not across networks, reducing the chances of being able to ask for assistance from other service providers. The occurrence of technical problems to such a communication path affects information sharing with other departments.

Information systems used in public safety networks require high standards of relia-bility and resilience to provide linkage and communicate easily to guarantee capacity in daily operations. Devices developed for use in emergency response and public safety require consistency across networks, devices, and applications. Such qualities ease the process of collecting, analyzing, and disseminating information to concerned agencies.

The adoption of new technologies into public safety networks will enhance interoper-ability between different agencies (Adam et al. 2008). Innovative technology will allow the adoption of a single cohesive unit where all communications reach all the agencies at the same time. Innovative communication channels will allow information sharing without barriers. The creation of a single cohesive wireless network for specific use by all the public safety and response teams would enable ad hoc information sharing without barriers to communication. It would allow sharing of physical and virtual resources across agents through the development of heterogeneous devices. A wireless network of information sharing eliminates the need for specific devices in communication. It increases trust among the stakeholders in emergency response and public safety. The members of different groups will contain unlimited access to information shared across networks. Shared networks enhance interoperability by allowing control over available resources thus tackling the issues in information and resource sharing.

Interoperability to enhance public safety during emergencies requires cooperation among different stakeholders and their emergency response units toward resolving common goals (Bedford and Faust 2010). The process requires information and knowledge sharing among themselves and the public affected by the occurrence of disastrous situations to deeper understand response channels. The process through which information is shared should be a priority defined before the occurrence of the situations requiring coordination of diverse participants. Collaboration increases the probability of achieving the goals of averting risks posed to human life through coordinated actions. It allows information sharing between the source and users through a defined communication channel.

Existing frameworks on emergency response are important in developing an interoperable system as they enable identification of gap areas (Adam et al. 2008). Established frameworks allow evaluation of the analysis of public safety based on existing communication channels. Despite the notable advantages that could be achieved from the interoperability of public safety through the effective exchange of valuable information, complete coordination remains a mere desire. However, the benefits accruing to the stakeholders are more if the information gap is reduced.

Existing frameworks point to several gap areas (Federal Communications Commission 2010). First, each agency utilizes its individual infrastructural facilities to enable communication among its employees and facilitators. The costs of building and maintaining such structures are high compared to if the different agency resources are pulled together to create a compatible system of emergency response. The agencies deploy a number of antenna resources to facilitate single agency information sharing that are bound to fail rather than combining the very same resources for a single efficient communication network path. Pooling resources would enhance interoperability between emergency response agencies by establishing a nationwide cellular network with enhanced functionality and coverage.

Secondly, economies of scale will result from the expected increase of network reuse by the emergency response and public safety networks (Bedford and Faust 2010). Notably, when an agency creates its individual channel of information sharing, the frequency of use is minimal. Fragmentation of systems into smaller units reduces efficient use of available resources. Interoperability will require a combination of resources to a common pool to create a single unique system that permeates the use of cellular services. Use of a single communication channel to share information will allow efficient use of the frequency. In a single agency system, minimum allocation of frequencies is larger, creating spectrum inefficiencies.

With such evident inefficiencies, the cost of operating a single agent communication channel is high compared to a shared network of communication (Peha 2007). High costs are driven by the small volume of usage of the communication network created for a single public safety network. Additionally, the cost of acquiring the public safety equipment is high for each line of public safety agency compared to if the resources are shared. Costs are further affected by market fragmentation and monopolistic suppliers. Combining efforts in the acquisition of the infrastructure used for public safety will grant the public safety agencies bargaining power and negotiating strength for development of an infrastructure based on compatibility to market needs (Federal Communications Commission 2010).

Employment of a broadband service across departments and agencies will increase interoperability, and dependency among safety networks as information will pass through a single network path (Adam et al. 2008). As a result, sharing knowledge and information will occur quickly without hindrances of physical or structural forms. Development of a single network of information sharing will occur at lower cost and enhanced capacity as it becomes possible to use existing commercial technology and infrastructure. A cohesive nationwide broadband for public safety agencies will increase interoperability by eliminating restrictions of access on others' networks due to lack of trust.

Public safety agencies face problems of interoperability since they contain overlapping functions amounting to continuous conflicts and lack of trust as each agent exists in competition with the other (Peha 2007). Jurisdictional power affects cooperation between the various agents as each attempt to prove prowess. Another impedance to public safety agencies acting in cooperation occurs from decentralization of functions. A lack of a unitary command system leads to conflicting interests as each department feels their area of operation should operate independently from the rest of the agencies (Federal Communications Commission 2010). The high number of agencies each with their unique infrastructure makes cooperation across fields impossible. Interoperability in such circumstances becomes near impossible due to differences in network range use.

The different public safety networks apparently use incompatible software and lack of logical inference makes interoperability difficult in the circumstances (Adam et al. 2008). Public safety networks use varying radio frequencies to avoid interference from other agencies. The varying range is a cause of the absence of interoperability since it is difficult to make a communication to another channel with a higher frequency. Efforts to establish systems with interoperability face difficulties due to the limitation of the system to voice sharing rather than data on possible threat areas. The existing gap in information sharing from infrastructural differences makes it almost impossible for collaboration between different public safety networks.

Interoperability between the public safety networks could face challenges from lack of supportive policies to ensure harmonization of efforts (Federal Communications Commission 2010). The majority of players feel the implementation of a cohesive broadband network to allow information sharing is expensive to install. However, the costs of fragmented communication are significant, making the establishment of joint public safety equipment a necessity. Consequently, the establishment of a system that allows collaboration between various agents, jurisdictions, and departments in the enhancement of a public safety network will reduce the loss of lives and property burden on the government.

Given the analysis, interoperability remains a key section of enhancing public safety measures by allowing cooperation between various agents concerned with emergency response (Bedford and Faust 2010). Public safety systems require access of information from diverse platforms to enable timely collaboration in averting extensive risks posed by disasters, natural or man-made. A unified communication system for public safety networks is essential to allow structured command, and resource sharing across geographical and department jurisdictions (Peha 2007). Increased access to such a system will allow decentralization of functions, command and control, efficiency in data sharing, and a quicker flow of information in the industry.

2.3.3 Benefits of Evolution

Interoperability is essential since it will allow the public safety agencies to share in the costs of handling emergencies, leading to economies of scale in handling an emergency. Use of innovative technology devices leads to information sharing without hindrances (Bedford and Faust 2010). Reduction of physical feature barriers will enable common planning of operations across agencies. The agencies will be in a position to utilize assistance from other players in the field. The various departments will be in a position to share intelligence and coordinate plans for a joint operation.

Transitioning to advanced communication equipment allows the public safety network to increase their geographical coverage (Bedford and Faust 2010). Increased information sharing eliminates jurisdictional restriction, allowing the initial investment to be recouped as the market expands. The costs of establishing a modern communication broadband network will be shared across public safety networks and spread to an extensive geographical coverage.

Establishment of a shared broadband network will allow increased information sharing by undermining existing disparities of jurisdiction and overlapping functions (Bedford and Faust 2010). A shared network will eliminate unnecessary cooperation and allow resource sharing as communication between agents is enhanced. Seeking assistance will take less time and the first responders will receive adequate training on system use and cooperation. The public safety agents will receive training on the utilization of systems, equally enhancing their interpretation of available data. Increased knowledge levels for the users of a shared network increases interoperability through raised interpretation of presented data.

Interoperability through a shared network will enhance equipment cost reduction (Peha 2007). Notably, sharing a broad network will allow standardization of equipment, making mass production easy. In contrast, the shared networks would allow the use of normal cellular equipment as opposed to the highly expensive equipment used in individual public safety networks. On the same note, use of standardized equipment will reduce dependence on public funding that takes a long time to receive. High reliance on unreliable public funding for equipment purchase is eliminated leading to higher rates of inter-agency connectivity from standard apparatus. Standardization of communication facilities allows ease of switching equipment, therefore adoption of new technological innovations that enhance connectivity will be simplified as opposed to having similar equipment due to fear of high costs of investment in modern equipment.

Creating a large extensive market for a range of equipment leads to cost-effective technology through shared research and development processes (Peha 2007). The engineering and structural facilities used in the development of communication equipment are shared. Growth in the market will lower costs of production enabling uniformity in infrastructure purchase at lower initial costs. The low-cost strategy ensures interoperability through the creation of standardized equipment affordable to all public safety network agents.

Moreover, deployment of a public safety broadband capable of utilizing commercial technologies and network design allows increased communication without imposed restrictions (Bedford and Faust 2010). Usage of widely established consumer technology will allow the use of industry innovations to share information across the network without fear of data loss and manipulation. Subsequently, the devices will require specialization to public safety needs to increase coverage to areas where commercial providers lack interest while maintaining their use of existing technologies. Remarkably, the extensive coverage of communication networks as more innovations hit the market will simplify information sharing as events take place (Bedford and Faust 2010). Also, the jurisdictional boundaries and trust issues will be eliminated. The standards of operation will be improved beyond commercial standards. Nevertheless, ensuring the networks continue to meet public safety standards year after year as technology and public safety needs evolve is a necessity.

Interoperability will be realized when a nationwide network is developed with the specification of public safety agency requirements (Federal Communications Commission 2010). The coverage, reliability, and security features should remain a paramount area. Enabling roaming on the shared network will allow information sharing even when congestion in the primary infrastructure is noted or when the infrastructure is destroyed by calamities. A multi-faceted approach to shared infrastructure enhances collaboration among first responders without creating issues of trust or inconsistency in network coverage at lower costs.

2.4 Conclusion

It is definite that the means of human communication has significantly enhanced throughout the years. Innovative technologies in communication have made it possible to accomplish connection between individuals and businesses by changing the way in which people interact on an everyday basis. Wireless communication systems are punctuated with important trends. These trends include the coming together of wireless and internet broadband communications and rapid developments as a result of large-scale circuit integration, digital signal processing, digital switching techniques, and digital and RF circuit fabrication. In the mid-1990s, 2G wireless communication systems were set up branded by digital modulation and voice-centric services. In order to be able to contend in the up-to-date economy, most populations have realized that they are in need of internet access. The world as we see it now is in the middle of an inspiring growth and revolutionary use of mobile broadband. The history of broadband wireless systems can be connected back to the necessity to find an economical option to the present traditional wireline-access technologies. Constant growth and expansion of the internet need the overall network architecture to evolve. There are a number of factors that have been set as determinants to the evolution of broadband internet access: government involvement, performance appraisals, and technological advancements in terms of system architecture. When talking about the benefits of the broadband internet access, it is necessary to shed some light on how public safety and security benefits from the use of LTE and 5G standards. In essence, the broadband system has come a long way and its constant evolution is a promise of better internet access and use for the future.

Existing public safety networks lack the capacity to share information across first responders. As a result, they hinder the ability to save lives and costs in a crisis. Enhancing interoperability among public safety networks and first responders requires the establishment of rules and agreements on how the partners will interact. The presence of guidelines and policies on cooperation is needed to enable industry standard setting. Development of an infrastructure to facilitate interoperability requires a framework to ensure situation-specific scope and applicability. Interoperability occurs along political boundaries; with a need for political support, legal conditions to enable collaboration through policies, and the possibility of the presence of integration among public safety network agencies. Standard ways of utilizing data from the other partners in a meaningful way need to be developed. The coverage, reliability, and security features should remain of paramount concern in infrastructure development. The 2014 National Emergency Communications Plan in the United States is legislatively required to set timelines to establish interoperability. From 2008, nearly 90% of the set milestones have been achieved. The development of the plan has created new jobs in state and local capacities. Public safety agencies now have tactical plans and protocols to communicate throughout the nation during emergencies. The emergency response system will continue to evolve along with the post-analysis of threat responses and emergence of new technologies and updating of current technologies and procedures.

2.5 Recommendations

- There is need to create a next generation nationwide public safety wireless network and adopt 4G cellular technology to leverage a fast pace of commercial development.
- Efforts should be made to address the growing digital inequality between developed and developing countries in terms of access to broadband internet access.
- Improvements should be made toward the quality of connection and "under-served" people. A mere 76% of the world's inhabitants live with access to a 3G signal, and only 43% with access to a 4G connection.
- Even though progress has been made, more should be done concerning affordability of broadband.
- Public safety networks and governments should focus adequate resources on investments in ICT infrastructure.
- There is need to carry out a review and bring up to date regulatory frameworks for broadband.
- The respective national governments have the task of developing and improving national broadband plans.
- There is also a need for benchmarking of trends and developments in telecoms and ICT by developed countries that have successful ICT infrastructure to help develop countries that are still struggling with the idea.

- There is need to consider infrastructure sharing for policy makers wishing to consider open access approaches to infrastructure.
- There is an absolute need to integrate public safety networks together to realize a converged networks, services, and applications.

References

Adam, N.R., Atluri, V., and Chun, S.A. (2008). Secure information sharing and analysis for effective emergency management. In: *Proceedings of the 2008 International Conference on Digital Government Research, Montreal, Canada*, 407–408. https://dl.acm.org/citation.cfm?id=1367832.

Astély, D., Dahlman, E., Furuskär, A. et al. (2009). LTE: the evolution of mobile broadband. *IEEE Communications Magazine* 47: 44–51. Document.

Bedford, D., and Faust, L. (2010). Role of online communities in recent responses to disasters: Tsunami, China, Katrina, and Haiti. Pittsburgh, PA. https://onlinelibrary.wiley.com/doi/pdf/10.1002/meet.14504701207

Broadband Forum Technical Report, 2012, issue no.1 Broadband Forum https://www.broadband-forum.org/technical/download/TR-145.pdf

Casteele, J.E. (2010). *Understanding Broadband - Advantages and Disadvantages*. Article. New York: Bright Hub.

Costello, S. (2012). *China Mobile Hong Kong to Construct LTE-TDD Network*. Article. Hong Kong: TT Magazine.

DSL Forum 2007 DSL Forum Technical Report Issue no.1 Broadband Forum https://www.broadband-forum.org/download/TR-144.pdf

Ekström, H., Furuskär, A., Karlsson, J. et al. (2006). Technical solutions for the 3G long-term evolution. *IEEE Communication Magazine* 44: 38–45.

Ergen, M. (2009). *Mobile Broadband – Including WiMAX and LTE*. New York: Springer.

Federal Communications Commission, (2010). A Broadband Network Cost Model, A Basis for Public Funding Essential to Bringing Nationwide Interoperable Communications to America's First Responders, Washington DC: Federal Communications Commission [Available online: http://hraunfoss.fcc.gov/edocs_public/attachmatch/DOC-297709A1.pdf

GSMA, The Mobile Economy, 2019. GSMA https://www.gsmaintelligence.com/research/?file=b9a6e6202ee1d5f787cfebb95d3639c5&download

Khan, F. (2009). *LTE for 4G Mobile Broadband – Air Interface Technologies and Performance*. Cambridge: Cambridge University Press.

Kumar, A., Liu, Y., and Sengupta, J. (2010). Evolution of mobile wireless communication networks: 1G to 4G. *International Journal on Electronics & Communication Technology* 1 (1): 68–72.

Peha, J.M. (2007). How America's fragmented approach to public safety wastes spectrum and funding Proc. Telecommunications Policy Research Conference, Sept. 2005. Expanded version in. *Telecommunications Policy* 31 (10–11): 605–618.

Program, Office of Emergency Communications/Interoperable Communications Technical Assistance (2011). *Understanding Mobile Broadband for Public Safety*. Periodical. New York: Homeland Security.

Sidak, G. (2016). What aggregate royalty do manufacturers of mobile phones pay to license standard-essential patents. *The Criterion Journal on Innovation* 1: 77–81.

Taylor, J. (2014). *Optus to Launch TD-LTE 4G Network in Canberra*. Periodical. Canberra: ZDNet.

The Broadband Forum (2014). *Multi-Service Broadband Network Architecture*. Fermont, CA: The Broadband Forum.

Williams, C.B., Fedorowicz, J., Sawyer, S. et al. (2009). The formation of inter-organizational information sharing networks in public safety: cartographic insights on rational choice and institutional explanations. *Information Polity* 14: 13–29.

3

Public Safety Communication Evolution

3.1 Introduction

3.1.1 Public Safety Network and Emergency Communication Networks

Contemporary societies are constantly grappling with public safety issues such as terrorism and violence, as well as disasters within their communities. As such, law enforcement agencies and other first responder agencies are keen to establish communication networks that will aid in fast response to contain the public safety problem at hand as soon as is possible. To achieve this, most agencies have adopted public safety networks (PSNs), which are communication networks used in situations where a fast response is required due to an incident that risks the public safety of Americans. With the technological advancement today, the public safety agencies have upgraded their communications network to include gadgets such as hand-held computers and mobiles that have an internet connection, equipping responders with personal devices for fast communication, as well as video cameras, among other things. To delve deeper into this concept, this chapter will explore the gradual change in PSNs from the traditional approaches to communications, and examine the new technologies that have permeated this realm, as well as their advantages.

As mentioned before, society is changing, perhaps proportionately with the technological advancement today. In the PSNs and emergency communications discourse, there have also been significant and visible changes. Before the internet era, societies were still kept safe by these agencies, and the communication networks were limitedly effective. For instance, in emergency response, the control center would dispatch the first responders to the scene upon receiving the 911 calls, which would sometimes take time. Additionally, the control centers were also using devices largely based on time-division-multiplexing (TDM) technologies that have gradually been rendered obsolete in a world of wireless and ad hoc networking. TDM technology transmits and receives signals through a common signal path using various synchronized switches that are found on the end of the transmission lines.

Most communications were one way, limiting timely interactions between the first responders and the control center. Most of the traditional communication systems in PSNs were based on wireline data connections, whose major impediment was their limited data capacity, which was unable to handle communications over a vast geographical area. Generally, the traditional PSN relied heavily on two-way communication, with 911 caller ID not visible to the 911 responders at the control center, hence relying heavily on

Public Safety Networks from LTE to 5G, First Edition. Abdulrahman Yarali.
© 2020 John Wiley & Sons Ltd. Published 2020 by John Wiley & Sons Ltd.

the caller to reveal their location as well as details of their concerns. Such inconveniences could cost lives in extremely dire situations due to poor communication services. However, PSNs are currently undergoing an upgrade in terms of software and hardware, which has increased their efficiency in enhancing public safety. This will be discussed in the next section.

The public safety entities are charged with the responsibilities of safeguarding and protecting the wellbeing of the general population in the case of man-made and natural disasters or eventualities. They undertake planning and preparation, and respond to emergencies (Bader et al. 2017). The emergency management actors include firefighting units, emergency medical services (EMS), law enforcers, rescue squads, and others, that will arrive first at the emergency scene. Thus, they are mostly referred to as emergency first responders (EFRs) (Baldini et al. 2012). The ability of such teams to communicate one-to-one, one-to-many or many-to-many in a seamless manner is essential for sharing critical facts that influence their ability to save people's lives or ensure their wellbeing.

Public safety communication has undergone significant advancements intended to strengthen national preparedness and enhance emergency communication abilities (Ferrus and Sallent 2015). From the earliest land mobile radio system (LMRS) critical communication systems that involved narrowband data during its release have progressed to wideband data such as the tetra enhanced data service (TEDS) (Baldini et al. 2012). Wideband networks provide massive data rates of several hundred kbps intended to meet public safety needs. While offering essential public safety communication like group messaging, calls, and voices, further evolution is required to achieve data-rich broadband services that support numerous devices, services, and applications to improve the operational capability of public safety communications (Bader et al. 2017). Some of the examples of the current public safety communication capabilities include remote healthcare, observation of the first responders, and video streaming. In contrast to the existing mobile applications in the market, public safety has stringent requirements regarding signal quality, coverage, ratability, and security (Ferrus and Sallent 2015). The development of future public safety communications needs to consider such requirements.

In the recent past, significant advancements have been witnessed, and are still underway, that have concentrated on the deployment of cutting-edge signal processing methods. The methods utilize dynamic multicarrier waveforms in public safety communication bands to fulfill the evolving novel data service requirements while working with the traditional networks within a similar frequency band (Baldini et al. 2012). The intent is to facilitate an effective transition to a nationwide broadband system and enhance spectrum capacity and efficiency. For instance, spectrum sharing is expected to offer considerable advantages while using the radio spectrum and assisting in mitigating the inadequate spectrum capacity problem experienced in public safety communication. Nonetheless, there are several hurdles that question that implementation of public safety communication systems and the incorporation with fifth generation (5G) communication (Fantacci et al. 2016). For example, integrating public safety communication system with Long Term Evolution-Advanced (LTE-A) are not augmented for mission-critical voice communications, particularly for groups (Bader et al. 2017). Consequently, there is a need for the traditional and emerging technologies in safety communication to coexist as amicable solutions are being researched.

The focus of this chapter is to reveal the evolution of public safety communication. The chapter takes into consideration the past and present technologies as well as the emerging and future development in public safety communication.

3.2 Public Safety Standardization

Standardization involves the process of developing and executing technical standards guided by agreements of various parties such as governments, users, standards associations, and interested groups. The goal of standardization is to enhance safety, quality, compatibility, repeatability and, ultimately, interoperability (Carla et al. 2016). Additionally, standardization makes sure that products or services adhere to legislation and sufficient levels of security. Between 2013 and 2014, there were significant and differing debates concerning the superior standardization for the future public safety technologies. After deliberations, it was agreed in 2014 that a standardization process would be carried out by 3GPP in a manner that incorporated public safety functionalities in subsequent releases of 3GPP (Chen and Mitchell 2014).

There were two major waves of evolution following the third generation (3G) of wireless mobile communications systems. Long Term Evolution (LTE) was defined in Releases 8 and 9 while LTE-A is defined within Releases 10 and 11 (Lin et al. 2014). Releases 12 and 13 are considered to be the first to address public safety requirements specifically as well as Releases 14 and 15 (Lundgren and McMakin 2018). Standardization of the 5G network and the Internet of Things (IoT) is still underway with standalone and non-standalone (NSA) specifications released in 2018. As compared to 5G, public safety standardization has not been accorded due priority. It is expected in the future that 3GPP will provide comprehensive standards and requirements for public safety systems (Lundgren and McMakin 2018). Table 3.1 summarizes the major standardizations in public safety.

3.3 Evolution of Public Safety Communication

The conceptual framework of evolution depicted by Figure 3.1 below reflects the deployment of wireless broadband communication while still retaining the land mobile radio (LMR) network to support the mission-critical communication via voice (Fantacci et al. 2016).

From the current state of affairs, LMR networks, the countrywide public safety broadband networks, and commercial broadband networks are evolving in parallel (Lundgren and McMakin 2018). Therefore, as communications transition, public safety will deploy the dependable mission-critical communications through the conventional LMR approach and, at the same time, safety entities will start implementing the embryonic wireless applications and services provided by broadband network connections (Ahmadi 2018). Throughout the changeover period, the public safety agencies should start developing their dedicated wireless broadband network to move away from the commercial broadband services. When the non-technical, as well as the technical requirements, are met, the agencies will entirely migrate and accomplish mission-critical capability (Blind 2016). Because wireless broadband technologies do

Table 3.1 Public safety standardization.

Standard	Features
Group Call System Enabler (GCSE)	The standardization involves a collection of various mechanisms providing multicast and unicast transmissions for group communication. It entails procedures for flood control and pre-emotion required by public safety (Carla et al. 2016).
Proximity Services (ProSe)	The standardization defines the architecture as well as radio interface used for direct device-to-device communications.
Isolated E-TRAN Operations for Public Safety (IOPS)	It includes service availability standardizations for public safety in delivering services autonomously when connections fail between the core network and base station. The standard targets at defining equivalent functionalities for LTE networks and it is the recommended standard for interface and protocols (Lin et al. 2014).
Mission Critical Services (MCS)	The applications utilized in mission-critical communications in public safety (such as MCVideo, MCPTT, and MCData) require a common array of capabilities to provide services to the users. Therefore, MCS standardization involves system capabilities specifications.
Mission Critical Push-To-Talk (MCPTT)	The standardization item describes the applications required to provide voice service for public safety communication.
Mission Critical Data (MCData)	The standardization specifies applications that provide non-real-time data services in public safety settings including text and multimedia messaging.
Interconnection of MC Systems	This item enables various MC systems to be linked to foster communications for mutual assistance eventualities.
Mission Critical Video (MCVideo)	Defines the applications delivering real-time video services within public safety.
Interworking with legacy PMR Systems	According to the standard, narrowband public safety will be deployed side-by-side with broadband.

not at present support the mission-critical voice feature (direct/talk around/simplex modes), there is a major epoch looming whereby LMR and wireless broadband will work together. In the past, communication regarding mission-critical voice has been delivered by employing LMR systems that achieve public safety standards and are operated by a single jurisdiction or agency (Lundgren and McMakin 2018).

3.3.1 Mission-Critical Voice

In daily operations, public safety actors require dependable voice communications for daily operations tactical situations and even for large-scale responses. Lundgren and McMakin (2018) note that voice communication offers the emergency responders an instantaneous, continuous and reliable connectivity between the responders and dispatch actors as well as amongst various responders. Therefore, the mission-critical voice is accomplished by employing dedicated LMR networks (Ahmadi 2018). It

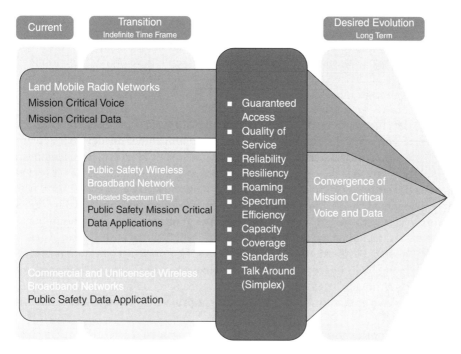

Source: Homeland Security, "Public Safety Communications Evolution", Brochure,
Nov 2011

Figure 3.1 The public safety communication evolution process. Source: Lundgren and McMakin
(2018).

ensures responder-to-responder and one-to-many responders' communications, which
is a vital attribute.

3.3.2   Mission-Critical Data

First responder entities utilize inadequate data communications for complementing
mission-critical voice (Fantacci et al. 2016). Currently, they employ data services for
basic functions like checking vehicle licenses, transmitting low-resolution images,
digital dispatch, and wanted individual queries. The emergency responders have
accomplished wireless data capabilities by using a commercial wireless service or building their systems (Blind 2016). While they are functional, data services are increasingly
inadequate in promptness/speed and do not offer real-time and advanced applications
required by the public safety communicators.

The public safety agencies foresee a dedicated network created to meet safety necessities that employ dedicated continuum. Thus, public safety actors have recognized LTE
as the standard tools for creating this network (Fantacci et al. 2016). Since this capability
is developed by utilizing LTE, public safety actors will not relent in working with government and other industries to advance the technology as well as address the requirements
essential for its deployment. In the transition era, public safety will start deploying LTE
for data applications in mission-critical communications (Lee et al. 2016s).

3.3.3 Requirements for Evolution in Communications

As mentioned earlier, there are several requirements that need to be met for effective evolution: funding, governance, resiliency, research and development, guaranteed access, roaming, quality of service (QoS), standards definitions, reliability, and talk around/direct/simplex mode (Lee et al. 2016).

- Funding. The emergency response actors face a big hurdle in supporting the development of mission-critical systems while intending to implement the developing technologies together with wireless broadband (Bonneville et al. 2016). To accomplish the evolution process, they require local, federal, or state funding to pay for various costs linked to researching and developing new and sustainable broadband networks (Bader et al. 2017). The Office of Emergency Communication (OEC) has been working to ensure the federal government provides grants to emergency communications agencies, including a coordinated grant that will assist in ensuring consistency in federal grants. Furthermore, OEC intends to seek funds that promote technical standards being built, interoperability being achieved and computability via coordinated efforts (Ahmadi 2018). Accordingly, the coordinated efforts will make sure all programs receive grants to support improvement projects in emergency communications.
- Governance. Coordination and cooperation across various agencies require successful governance structures (Lundgren and McMakin 2018). A countrywide interoperable wireless broadband for public safety communications needs a countrywide structure of governance. Besides technical and managerial skills, governance calls for active engagement of emergency communication stakeholders who perform duties at the tribal, national, local, and state levels through various disciplines and jurisdictions (Bader et al. 2017). Still, for governance, planning policy and partnership are vital aspects. Concerning planning, the public stakeholders should engage in countrywide, regional, and statewide tactical planning. The partnership involves the willingness of various jurisdictions and disciplines to partner together and develop compatible solutions. As the agencies strive to create more comprehensive and reliable solutions, the partnership is considered vital, especially in creating open standards and nationwide networks that meet the needs of entire nations (Bonneville et al. 2016). Moreover, it is essential that all levels of government and agencies proactively and collectively create policies as well as strategies concerning emergency communications.
- Research and Development. R&D efforts are vital in making sure that emergency communications achieve effective, interoperable, reliable, and standardized capabilities (Thomas and Squirini 2018). They will foster the establishment of new systems and test how such systems will work to meet the requirements of public safety communication (Lundgren and McMakin 2018). Additionally, R&D will enhance sustainable functionality during harsh conditions where the emergency teams work.
- Access. The public safety actors should have assured access to instant and reliable communication capabilities at all times, 24/7, to respond effectively (Ohtsuji et al. 2018). Guaranteed access is a significant attribute in ensuring public safety, particularly when utilizing broadband networks.
- QoS. Safety communications calls for systems and networks that can assure a specific level of performance for the critical systems. QoS defines how specific types of data

are handled and their prioritization among applications and users. It ensures reliable performance (Bader et al. 2017).

- Reliability. Reliability is an essential technical requirement of safety communications that ensure emergency actors can rely on the systems and networks for mission-critical communications (Ohtsuji et al. 2018). That is, networks and systems need to be developed in a manner that reduces service degradation or minimizes the loss of capacity.
- Resiliency. Networks and systems supporting public safety response and communications need to be resilient including high reliability, availability, redundant power, communication paths, components, and infrastructure (Thomas and Squirini 2018). This will minimize service disruptions.
- Roaming. For the communication systems and networks to undertake their roles efficiently, the responders should be allowed to seamlessly roam between the commercial and PSNs when needed (Thomas and Squirini 2018).
- Efficiency and capacity. The high rate of progression of wireless technologies and broadband enabled services and applications is placing limitations on the available capacity spectrum in the marketplace. This has rendered the networks unresponsive or slow. As such, this impacts negatively on public safety communications and response, which need 24/7 access to information to accomplish their missions (Su et al. 2018). There is a need for sufficient capacity as well as efficiency if the country-wide PSNs are to fulfill the emergency response requirements. Additionally, public safety needs a wireless signal coverage that makes sure reliability in operations in massive geographical areas like major population centers or rural areas.
- Standards. Technical standards are vital for interoperability. The standard based systems offer backward computability that enables actors to continue communicating via their mission-critical voice systems (Thomas and Squirini 2018). The LTE standards utilized by the private sectors are increasingly endorsed by the Federal Communication Commission to be deployed in the next generation networks (Thomas and Squirini 2018).
- Talk around/direct/simplex modes. Direct mode, also called simplex or talk around, involves the ability to talk device-to-device (D2D). It is considered to be significant since it enables response groups in operations to converse directly to one another while outside the coverage of network infrastructure (Su et al. 2018).

3.4 Public Safety Networks

PSNs involve a wireless network deployed for emergency services' communications like in EMS, police, and fire departments (Kumbhar and Güvenç 2015). A PSN can span a large borough with the critical data been relayed by broadband wireless network based on mesh topology like LTE or Wi-Fi (Rizzo 2011). Additionally, the PSN can be linked to mobile computing applications to augment the efficiency of emergency leading responders as well as the wellbeing of the public. The study broadly classifies PSNs into two classes including broadband networks and LMRS where TETRA (Terrestrial Trunked Radio) and APCO-25 standards belong to LMRS (Chow et al. 2012). On the other hand, LTE (FirstNet Broadband PSC) networks belong to the broadband network (Iapichino et al. 2014). The evolution of PSNs is summarized in Figure 3.2

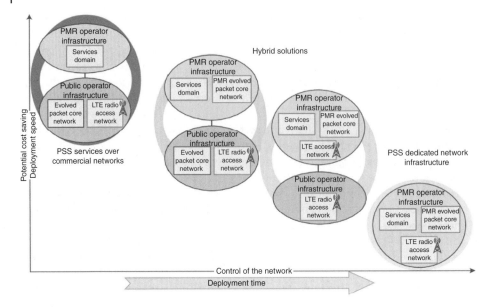

Figure 3.2 Evolution of public safety networks.

(https://www.semanticscholar.org/paper/Public-safety-networks-evolution-toward-broadband%3A-Fantacci-Gei/c124be6cf5127e0b5b77738945a20862db6f6e3b).

3.4.1 Land Mobile Radio Systems (LMRS)

In contemporary times, public safety communications are deploying the conventional LMR to support communications in mission-critical operations (Kumbhar et al. 2017). LMR is also known as public radio mobile radio involving person-to-person communication through voice communication comprising two-way radio transceivers that are mobile and portable and can be installed in vehicles (Kumbhar and Güvenç 2015). LMRS equipment is depicted in Figure 3.3.

The systems deploy UHF and VHF bands hence a limited range of between 4.8 and 32 km depending on the geographical terrain (Martinez and Vaughan 2012). However, repeated installation on mountain peaks, tall constructions, and hills can be utilized to enhance the coverage, as depicted in Figure 3.4.

The older systems used FM or AM modulation, but the most recent systems deploy digital modulation enabling them to transmit data and sound (Martinez and Vaughan 2012). The main objectives of the LMRS are to offer mission-critical communications to allow cohesive data and voice communication during emergency response and ensure reliability and interoperability as well as ruggedness (Paulson and Schwengler 2013). LMRs are perceived to offer reliable approaches for actors in the field to relay messages or connect with one another and with control or command centers (Kumbhar et al. 2017). As the LMRS have advanced over several decades, there are numerous systems that are incompatible but still utilized in public safety communication. Hence, the first responders are increasingly restricted by the fragmented networks and several decades-old wireless infrastructures. The public safety field still struggles to communicate effectively across various agencies and jurisdictions (Paulson and Schwengler 2013).

Figure 3.3 Two-way communication LMRS. Source: https://trollingpowersolution.com/how-to-get-fcc-license-for-two-way-radio.

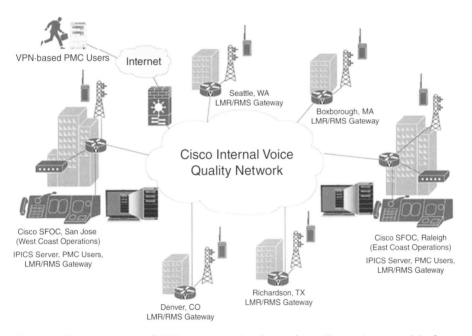

Figure 3.4 The components of LMRS communication. Source: https://www.picswe.com/pics/lmr-land-f1.html.

Additionally, public safety organs deploy a mixture of high-speed broadband data as well as low bandwidth involving commercial networks to support the response efforts and in undertaking functions like license plate queries, transmitting low-resolution images, text messaging, and digital dispatches (Kumbhar and Güvenç 2015). While such advances are considered worthwhile, most of the present solutions of broadband data have the inadequate capability to support emergency responses due to lack of interoperability or are not built to meet the public safety standards (Martinez and Vaughan 2012). On the other hand, that much anticipated wireless broadband cannot currently meet the public safety communication requirements for emergency responders (Kumbhar et al. 2017). Therefore, LMR will be around for several years and can only be complement by wireless broadband.

3.4.1.1 SAFECOM Interoperability Continuum

The five important aspects outline in SAFECOM interoperability band involve governance, technology, usage, standard operating procedures and training, and exercise. These aspects apply to LMR communication planning as well as broadband (Sun et al. 2015). Hence, they should be considered when planning to deploy interoperability solutions in safety communication (Thomas and Squirini 2018). The continuum will continue being deployed in guiding interoperability planning and frameworks.

3.4.1.2 Wireless Broadband

Wireless broadband offers rapidity in data communications, particularly in the mobile context. Presently, safety communication utilizes broadband for data applications but not in mission-critical voice communication for first responders (Baldini et al. 2012). While data applications will not offer LMR voice capabilities, such technologies are considered important and can significantly enhance emergency responses (Baldini et al. 2014). Since public safety is a unique mission, the emergency responders need to be equipped with wireless broadband devices as well as services that will assure access, coverage, and security as well as a high level of reliability, which is unlikely to be provided by commercial entities (Li et al. 2015).

It is believed that utilizing an economical countrywide public safety wireless broadband network will offer public safety agencies access to more advanced, exceptional, and cutting-edge applications and technologies that can enhance emergency response activities (Baldini et al. 2014). This will ensure access to real-time multimedia information that will drastically enhance communication ability and access to information required for informed decision making. For instance, public safety entities can easily access video images involving criminal activities while in progress, download the master plan of buildings that are on fire to handheld devices, or connect securely and quickly with persons from other cities and towns (Li et al. 2015). In a similar manner that smartphones have significantly altered the manner in which businesses operate, countrywide wireless broadband can massively change the communications and operations of emergency responders (Ferrus et al. 2013a, b).

The recent advances regarding wireless data communications have led to increased access to applications and delivering instantaneous information required by public safety agencies (Fantacci et al. 2016). Whether deployed on a daily basis or large-scale responses, such technologies can enhance communications in emergency events and improve the effectiveness of responses. For instance, Advanced Automatic Crash Notification illustrates that extended wireless broadband applications can offer instantaneous useful information to the emergency response teams for decision making (Li et al. 2013). With the right information (photos, weather reports, prison records, drivers' photos, and licenses) being delivered to the right people at the right time, effective responses will be achieved (Baldini et al. 2012). Such applications and others can only be achievable via high-velocity wireless broadband support.

While the public safety agencies have for a long time recognized the significance of such developments, they are also aware that several hurdles need to be overcome and the requirements of such technologies meet the standards of other communications (Li et al. 2015). Therefore, it will take some time before such requirements are met and wireless broadband is incorporated into public safety activities. Also, such technology needs to be integrated and aligned with the commercial setting out of the LTE wireless services to be on a par with emerging technologies and take advantage of their cost efficiencies (Ferrus et al. 2013a, b). Future public safety communication ought to deliver mission-critical data and voice communications to various public safety actors in local, tribal, and state jurisdictions throughout the country and offer federal responders and others secondary users (like utilities and transportation) advanced capabilities (Ferrus et al. 2013a, b). Nevertheless, transitioning from LMR to broadband wireless networks will not be achieved overnight. The federal, private, state, and local actors need to work together in developing standards and requirements for assuring mission-critical operation communication.

3.4.1.3 Wi-Fi in Ambulances

It is necessary to note the important improvements made in ambulances in contemporary EMS communications networks. Ambulances are now fitted with computers that have wireless internet, which is mostly ad hoc in nature, that communicates with the hospital, sending patient information to the hospital staff who in turn prepare for their arrival adequately. In most cases, the ambulances are fitted with mobile gateways that provide an ad hoc network connection for the machines in the ambulance, and hence help to send real-time information to the hospital. Additionally, the Wi-Fi networks in the ambulances are critical in locating the emergency callers once the emergency control center dispatches the teams and the responders. Sometimes locating the caller may be difficult, but with an internet connection the first responders are able to use online map services to locate the callers. While traditional ambulances may have had inbuilt GPS systems, the online maps accessed through applications such as Google Maps are more efficient and accurate in locating a person, perhaps saving a life.

3.4.1.4 Satellite Communications in EMS and Public Protection and Disaster Relief PPDR

Sometimes there are situations where natural disasters occur. To enhance disaster preparedness, satellite communications are necessary for alerting the various disaster management agencies to take the necessary action such as evacuating people and sending out warnings to ambulances and other public safety organizations through the internet and other forms of connectivity. Currently, most satellite communications are using the TETRA digital standard, which was defined by the European Telecommunications Standards Institute (ETSI). TETRA operates on the Time Division Multiple Access (TDMA) network, which is critical in the provision of cellular communications. A major advantage of the TETRA network is that it allows for the transmission of voice and data in a variety of ways, and can operate in a direct mode, making it suitable for use as a walkie-talkie (Management Association, Information Resources 2013, p. 382). Additionally, TETRA covers a wider coverage zone, meaning that first responders across cities can communicate in the event of a natural disaster such as an earthquake, as part of the public protection and disaster relief (PPDR) efforts (Management Association, Information Resources).

Additionally, ambulances are now fitted with mobile satellite services which often use portable phones and terminals. For efficiency, the terminals are often placed on the ambulance for better reception, allowing the paramedics to communicate easily with on the ground observers as well as the callers. To add to this, an advantage of using the TETRA technology is that on its gadgets, TETRA provides more stable lines and a long signal range. However, a major setback attached to using the TETRA network for EMS and PSN communications is that the networks are slow and costly to maintain (Management Association, Information Resources). To add to this, the TETRA networks used in satellite communications are also considered a waste of spectrum, especially when used on radio. To enhance the efficiency of the satellite communications, most first response organizations will provide their employees with satellite phones.

The most important advantage of the satellite phone is that it is water and dust resistant. Additionally, the satellite phone allows the first responders to call any location globally, making communication easier. It is also advisable that organizations use the satellite phone in PPDR because it has fast voice and data services.

3.4.1.5 Technology in Patrol Communications

4G LTE networks have greatly evolved the speed of communications in the PSN arena, allowing for those, especially on foot, making ground observations in disaster areas, to communicate efficiently. As such, there are patrol solutions such as drones fitted with 4G LTE networks allowing for coverage over a vast area. The drone communicates with the mobile patrollers on their phones, and in the communication systems fitted in patrol vehicles in the area. Therefore, should there be any form of alert, the mobile patrollers performing ground observational services, using the 4G networks hosted in the drone, can communicate with patrollers in a patrol vehicle for further assistance. This mode of communication is efficient because not only does it cover a vast area, but it is effective in immediate and fast communication.

With time, these gadgets should be able to accept upgrades such as the 5G LTE, which would be faster and more efficient. This technology is mostly used by foot patrollers: personnel working in the forestry department, search and rescue responders, defense workers, and those in disaster response. For organizations working with these in a PSN, it is crucial that it invests in devices that can support the fastest internet generation, which is currently 5G. A major disadvantage of using PSN communication technology for communication and relaying of messages is that when the Wi-Fi connection is low the communication process is halted. The three major points of connection must be connected for the fast responders to experience fast communication. Therefore, in a situation where one of the recipients is disconnected, the communication process becomes obsolete.

3.4.1.6 Video Cameras

The use of CCTV in ambulances to record the occurrences both inside and outside the ambulances has increased with time. Using video cameras has become essential because of purposes such as record keeping, which is critical, especially in situations where there are court cases. The use of video cameras to record the proceedings inside the ambulances during an emergency is critical for purposes of accountability, especially when the proceedings and procedures are put into question. For instance, in situations where the patients files a claim that the paramedics sexually molested them, the video surveillance footage is a critical tool in helping clarify such issues and avoid unnecessary lawsuits. As such, it is necessary to equip emergency vehicles with functional cameras for accountability reasons. Additionally, the video cameras, especially in hospital facilities, have proved important not only in enhancing the security of the premises but also in promoting accountability among healthcare practitioners. For instance, CCTV footage provides information on what time a physician reported to work and checked out. Another advantage of introducing video surveillance in EMS ambulances is that it provides a platform for the paramedics and other medical personnel to chronicle the treatments used on a patient for future reference. It is a way of storing information. Lastly, video surveillance is also achieved through body cameras worn by the first responders, and as such they become effective in detailing to the PSN organization how the responders handled the situation, and how exactly the situation unfolded on the ground.

While video surveillance in the ambulances is critical for the aforementioned reasons, there are still concerns about the violations of privacy that this technology presents. This is because all forms of patient interaction are chronicled. This includes the fact that information on their health condition is disclosed and becomes susceptible to sharing, hence violating the patient's privacy. Due to the violation of privacy presented by the presence of the cameras, some patients may express concerns that they feel exposed and therefore refuse to disclose important health information, which could jeopardize their health outcome. While it is important that the introduction of video surveillance as part of PSNs and emergency communications, it is necessary that the various PSNs and emergency organizations strive to ensure patient privacy.

Contemporary PSNs and the emergency communications networks have evolved to adopt newer technologies that ease and hasten the communication processes in urgent situations, as well as in emergencies. The traditional approaches to PSNs were effective then, but now there is a need to introduce wireless connections and devices that are more efficient in the communication process, especially in this era of the internet.

3.4.2 Drivers of the Broadband Evolution

The major drivers of public safety communication to employ wireless broadband include:

i) Operational efficiency. The mission-critical services and operations of public safety are expected to advance to the next level of efficiency by utilizing new applications that are based on broadband data communication. The desired efficiencies are based on mobilizing the current office-bound applications and allowing entirely new applications (Baldini et al. 2012). Some of the examples that can improve efficiency among first responders include image transfer, video transfer, remote data access, and mobile offices.

ii) Safety improvement. In the recent past, various threats have emerged. For instance, terrorism is making the world insecure and many governments are on the move to utilize the most effective strategies to save lives once man-made or natural disasters occur (Ferrus et al. 2013a, b). Mission critical communications require a broadband connection like situational awareness software, and by utilizing broadband, the safety of citizens affected by risk as well as the first responders can significantly be improved (Li et al. 2013).

iii) Cost-effectiveness. The budgets of both government and public safety agencies (first responders) are being constantly reduced and so pressure-critical communications should contribute to cost savings (Fantacci et al. 2016). The cost related benefits of broadband networks involve a combination of enhanced efficiency and productivity, and new models of business such as day-one capital expenditure (CAPEX) investments can be substituted by using monthly operating expense (OPEX) fees (Baldini et al. 2012).

3.5   4G and 4G LTE

As wireless technologies have altered the way people live, it is time to rethink how they can affect positively public safety communication. 4G is known as the fourth generation of mobile phone technology that came after the third generation and second generation mobile networks (Dahlman et al. 2013). 4G is believed to build upon what 3G provides but does so in a rapid way due to its improved speed. 4G wireless is used to define the fourth generation of wireless cellular services (Dahlman et al. 2013). The potential applications of 4G involve high-definition mobile TV, mobile web access, 3D television, gaming services, and video conferencing. In 2008, the International Telecommunication Union-Radio (ITU-R) communication industry launched specifications for 4G standards that set out the peak speed requirements at 100 Mbps (12.5 MBps) (Cox 2012; Dahlman et al. 2013). This was meant to achieve high mobility communication in cars and trains and 1 Gbps for low mobility communications like stationery users or pedestrians.

While the majority of 4G services are branded 4G or 4G LTE, the attributes are not similar. Some deploy WiMAX technology in 4G networks and others use Verizon Wireless, resulting in a technology known as LTE. Gomez et al. (2014) observed that 4G WiMAX networks deliver speeds of 7–10 times faster than those of 3G, with speeds topping out at 10 Mbps. On the other hand, the Verizon LTE network offers speeds between 5 and 12 Mbps (Lien et al. 2016). LTE is perceived to offer superior speeds similar to the wireless broadband network. Efficient critical communications systems can consider the use of LTE to ensure the first responders communicate with others in a more reliable and rapid manner (Dahlman et al. 2013).

The National Public Safety Telecommunication Council (NPSTC), among other entities, acknowledged the desirability of achieving interoperable national standards for PSNs employing broadband capabilities (Abdelhadi and Clancy 2015). Thus, in 2009, NPSTC chose LTE as the technology for the national network and reserved spectrum of 700 MHz band specifically for LTE PSNs. In boosting research and development in the arena, they committed $7 billion in 2012 (Cox 2012). Huang et al. (2012) noted that LTE is one of the emerging technologies of choice for agencies to transform their mission-critical technologies. Lien et al. (2016) found that over 500 operators had commercially launched LTE, LTE-Advanced Pro and LTE-Advance in over 170 nations by 2016.

After a decision was made by NPSTC to adopt LTE, they began active engagement with the LTE standards community in 3GPP to develop the system enhancements required to meet public safety requirements (Abdelhadi and Clancy 2015). In the entire journey, NPSTC has shown strong aspirations that have been hailed by the commercial cellular communities (Shajaiah et al. 2014). The major factor of why LTE is growing in popularity involves its features that explicitly target mission-critical services. For instance, the prioritization of service features, as well as pre-emptive network access, enables an entity to continue working in case of disaster. It is a critical capability that makes it viable for mission-critical communications. Also, traffic of the concerned agencies can be prioritized and dedicated. Shajaiah et al. (2014) indicated that LTE can employ the existing service operators for the 4G/LTE network service, which eliminates the need for a dedicated frequency as well as availability problems.

LTE in public safety means a massive amount of data can be transferred instantly. This implies that high-quality images, as well as videos, can be conveyed, allowing the deployment of different applications like live streaming during an eventuality (Gomez et al. 2014). Similarly, mapping software can be utilized for location and even for direct-ing mission-critical resources to the right location. 4G LTE can also be used to unify the different PSNs being used. For instance, before a disaster response, actors use vary-ing frequencies and technologies like TETRA, VHF, and UHF. The different networks make inter-agency coordination, collaboration, and communication very challenging (Cox 2012). In Figure 3.5, we examine how numerous actors in public safety are unified via 4G LTE.

Rapid responses can be achieved when such networks are unified via LTE. Further-more, LTE satisfies rigorous requirements and standards for public safety communica-tions and networks like security, QoS, and resilience.

3.5.1 Benefits of 4G LTE in Public Safety Communication

The benefits of deploying 4G include improving upload and download speeds, clearer voice calls and communication, and reduced latency. According to Dahlman et al.

Figure 3.5 Public safety LTE configuration and unification. Source: Dahlman et al. (2013).

(2013), the standard 4G (or 4G LTE) has been found to be 7 times faster as compared to 3G networks and provides theoretical download or upload speeds reaching 150 Mbps. This compares to 80 Mbps maximum speeds in practice. Long Term Evolution-Advanced (LTE-A) can offer up to 300 Mbps (Shajaiah et al. 2014). Concerning latency, devices linked to a 4G mobile network gain a speedier response to requests made as compared to the previous generations of mobile networks. Huang et al. (2012) argue that the latency time has declined from 120 ms in 3G to about 75 ms for 4G. While it might not seem substantial in the study, it makes a significant difference when streaming live video from an emergency site (Kumbhar and Güvenç 2015). Additionally, 4G/LTE achieves superior voice communication clarity similar to VoIP (voice over internet protocol) that employs voice applications like Skype supporting voice over the internet. Voice over LTE (VoLTE) is utilized in 4G networks, achieving clear video and voice calls in 4G networks (Doumi et al. 2013).

4G LTE is known for providing instant communications during network outage that can be the case when a disaster occurs (Shajaiah et al. 2014). Strategies such as bring-your-own-coverage (BYOC) as well as network-in-box (NIB) can offer coverage at any location. NIB consumes fewer batteries and is considered to be a sustainable and practical option in emergencies (Doumi et al. 2013). A portable LTE utilized for public safety communication will enable connectivity between individuals and things resulting in better control of resources (Dahlman et al. 2013). The solution is particularly applicable when there is no connectivity in the area. Therefore, the first responders can configure their own broadband communication network and minimize the implications of a network outage that can affect rescue missions. Also, plug-n-play and the ease of usage can make sure that during a crisis, communication is easily established and maintained.

To achieve the needs of mission-critical requirements evident in public safety communication, 4G/4G LTE can be increasingly beneficial. The three areas of investment in 4G networks involve core networks, applications, and applications (Doumi et al. 2013). Concerning radio networks, it is important that 4G/LTE guarantees availability in mission-critical operations, including network coverage extended by advanced radio sites while the existing radio networks are overloaded. Rapid deployable 4G LTE systems can be utilized in extreme conditions with the radio network being shared by public safety as well as commercial users, regulated by access policy and QoS mechanisms (Doumi et al. 2013). There is a need to invest in dedicated applications in public safety such as push-to-video and push-to-talk that can improve communications among the first responders. As noted, another area of investment involves core networks. Dedicated network components are deployed at secure facilities. The virtualized core networks can also be deployed via flexible, dedicated gateways to ensure high security levels.

3.6 Fifth Generation (5G)

Fifth-generation (5G) wireless involves the emerging and state-of-the-art cellular network iteration that has been engineered to significantly enhance the responsiveness and speed of wireless technology (Hu 2016). 5G involves a wireless network comprising cell sites that are subdivided into partitions that convey data via radio waves. The 4G LTE wireless network offered the basis for the development of 5G. In contrast to 4G that calls for large and high-powered cell towers for them to transmit signals over an outsized geographical area, 5G signals are transmitted through several small cell stations positioned in an area, like light poles or on building roofs (Wei et al. 2017). It employs several small cells since the millimeter wave spectrum (a band of between 30 and 300 GHz) that fifth-generation depends on to produce high speed can only transmit over a short distance (Hu 2016).

The 2G, 3G, and 4G technologies employed lower-frequency bands of spectrum. To offset the millimeter wave problem concerning distance and interference, the network and wireless sector has considered the use of lower-frequency in relation to the fifth-generation network (Khan et al. 2017). Thus, network operators are likely to deploy the existing spectrum to develop their own novel networks. The lower frequency spectrum can reach longer distances but has the challenges of capacity and lower speed as compared to the millimeter wave.

The 5G network architecture is depicted in Figure 3.6.

Currently, 5G is utilized in four nations: China, the US, South Korea, and Japan, who are considered the drivers of the fifth-generation network. Wei et al. (2017) argued that network operators will use billions of dollars in the 5G network up to 2030.

Figure 3.7 demonstrates the adoption curve for the fifth-generation network.

There are two major services of 5G available: 5G cellular services and 5G fixed wireless broadband services (Fodor et al. 2014). The fixed wireless broadband services can provide internet access to businesses, enterprises, and homes without using a wired connection (Rouse et al. 2018). To achieve this, the network operators use new radios (NRs) in small cell sites close to buildings to beam signals to receivers on rooftops or windowsills where the signals are amplified. The fixed broadband services are anticipated to be cost-effective to operators when delivering broadband services to businesses since the approach avoids the requirements of rolling out fiber optic lines to each premise or

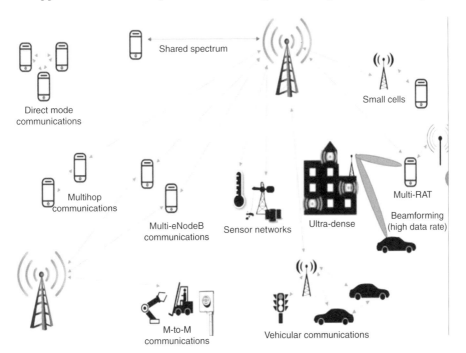

Figure 3.6 The architecture of 5G in public safety. Source: Wei et al. (2017).

2017
2018
2019
2020
2021
2022
2023
2024
2025
2026
2027
2028
2029
2030
Trials
Early adopters
Early majority
Mainstream adoption
Late majority
Laggards

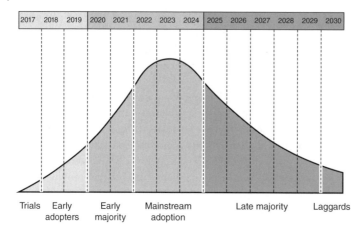

Figure 3.7 The adoption curve for the fifth-generation network. Source: Rouse et al. (2018).

dwelling place, instead, the network operators will set up fiber optics to the cell sites, and the users will obtain broadband services via wireless modems positioned on their premises (Hu 2016).

Regarding the 5G cellular services, they offers users access to the cellular network operators' 5G. The services are anticipated to be available by 2019 when 5G devices become commercially available (Rouse et al. 2018). Cellular services are dependent on the completion of 5G core standards in 2018, which has been realized. The fifth generation's first phase specification (Release 15) came into existence on March 2019 and will accommodate commercial deployments. Release 16 (second phase) will be completed by 2020 allowing submission to the International Telecommunication Union (ITU) (Rouse et al. 2018). The diagram in Figure 3.8 depicts the development of the fifth-generation network.

2017
2018
2019
2020
2021
2022
2023
2024
2025
2026
2027
2028
2029
2030
December 2017: 3GPP approves 5G NSA Standards
June 2018: 3GPP approves 5G SA Standards
4Q18: First 5G FWA commercial implementations occur
1H19: First 5G-compatible smartphones available
2019: First mobile 5G deployments using SA standards occur

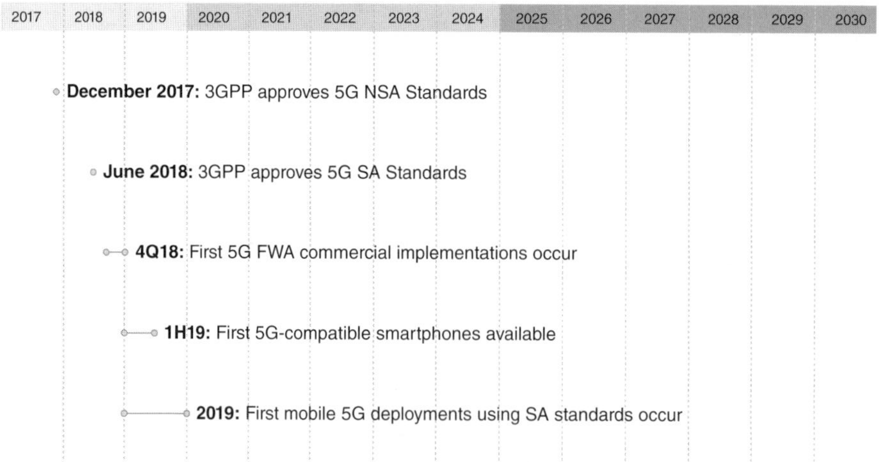

Figure 3.8 The development of 5G. Source: Rouse et al. (2018).

3.6.1 Performance Targets and Benefits of 5G

One of the essential requirements of the fifth-generation network involves address-ing the needs of professional and vertical markets. The vertical markets include IoT, mission-critical communications, automobile, and industrial solutions (Wei et al. 2017). Based on 5G, data can be transmitted over wireless broadband traveling at rates of up to 20 Gbps, which surpasses wired networks. It offers latency of 1 ms or lesser for users requiring real-time feedback (Ahmadi 2018). 5G will also allow a rapid escalation in the quantity of data that is transmitted over wireless networks, which is attributable to a greater bandwidth coupled with advanced antenna architecture (Williams 2017).

3.6.1.1 Security and Reliability

Some of the key requirements for public safety involve high availability, security, and reli-ability. When the first responders are in a dangerous situation, they need to rely on effec-tive radio coverage (Rodríguez and John Wiley & Sons 2015). 5G is perceived to offer numerous new capabilities that will enhance security, availability, and reliability of pub-lic safety communications. For instance, D2D communications, mobile-edge computing (MEC) and user-and-control plan separation achieved through software defined net-working (SDN) can provide exceptional capabilities (TCCA 2017). Additionally, the flex-ible deployment of radio services based on multi-connectivity technology will deliver superior reliability. Others aspects of 5G that are important to public safety include automated and connected cars that deliver real-time data while achieving ultra-high reliability, low latencies, and security (Naqvi et al. 2018).

3.6.1.2 Traffic Prioritization and Network Slicing

Mission-critical communications require networks that provide services in the worst-case scenario (Ahmadi 2018). Network capacity should consider events where a large number of people occupy a small area and network loading is at its peak. For instance, during a disaster, the emergency responders want to have high capacity while the rescuers and other teams want to communicate. In the case that they are employing a similar network, the priority is to enable the required capacity as well as the capacity for the emergency providers (Luo and Zhang 2016). It involves the capability to rigorously pre-empt some of the users so that services are offered to the first responders. The fifth generation is believed to add to the existing infrastructures allowing numerous users with varying requirements to operate within one network. Through network slicing technology, it is capable of subdividing one network into various virtual networks with varying priorities and use cases (Rodríguez and John Wiley & Sons 2015). Thus, network infrastructure can be shared for various niche operators including first responders. The separation of user and control plane traffic by employing network function virtualization (NFV) and SDN technology is imperative in the management of dynamically changing capacity requirements in a single network (TCCA 2017). The fifth-generation network offers novel capabilities for managing a portfolio of various use cases with varying priorities.

3.6.1.3 Facial Recognition and License Plate Scanning in 5G

Video capabilities like license recognition and facial recognition that are presently deployed are attributed to 4G technologies. Automatic number-plate recognition

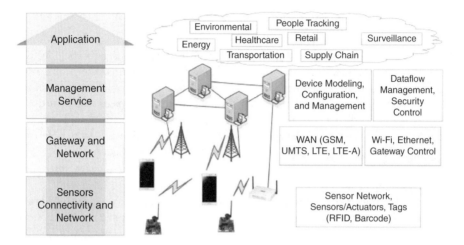

Figure 3.9 The proliferation of sensors and IoT. Source: https://opentechdiary.wordpress.com/2015/07/16/a-walk-through-internet-of-things-iot-basics-part-2.

(ANPR) utilizes optical character recognition of the image to read the registration plate of a vehicle and determine its location (Dahlman et al. 2018). Regarding facial recognition, this is a real-time technology that can help in disaster management and crime prevention (TCCA 2017). With gigabit 5G capabilities the first responders can take advantage of advanced facial recognition without requiring significant hardware upgrades as well as enhanced accuracy of tools deployed in reading the license plates of cars (Notwell 2018). The technology will translate to other IoT tools like tablets, cameras, and smartphones, which need a separate network for collecting and transmitting massive amounts of information.

3.6.1.4 Support for Sensor Proliferation and IoT

IoT is a major driver of the fifth-generation network supporting numerous machines, devices and sensors in communicating with one another, automating processes, and exchanging information (Dahlman et al. 2018), as shown in Figure 3.9. Some of the requirements of 5G include ultra-low-cost devices, high network capacity, and longer battery life. 5G provides a robust backbone for sensor proliferation, and it is expected that there will be a rise in IoT device adoption among the public safety communities as the novel sensor tools become available (TCCA 2017). For instance, the IoT sensor might be used by law enforcement vehicles to alert communication centers in case a shotgun is pulled from a police vehicle's rack. Sensors will provide information alerting first responders when a weapon is being used to enable them to tap into cameras to determine what is happening (Notwell 2018). Additionally, 5G will escalate the use of fingerprint sensors for identification of victims during an emergency or the criminals responsible for a crime. While such devices are increasingly linked with 4G routers, the advent of 5G will enhance message passing and analysis of data.

3.6.1.5 Reduction of Trips Back to the Station

One of the challenges that first responders and police face today involves the number of trips and traveling trips to and from their stations (Notwell 2018). The first responders and law enforcers need to be on the streets providing services to people and not

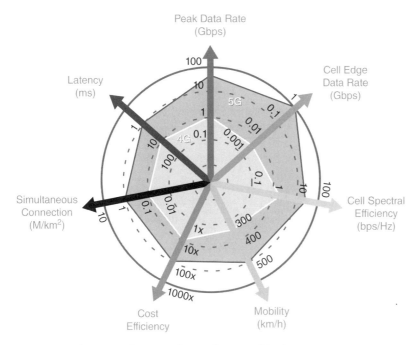

Figure 3.10 The major 5G content. Source: Samsung (2015).

filing reports at their stations. By using 5G- or LTE-enabled vehicles, the responders and police officers can easily access mission-critical data and applications to file reports from their vehicle, which will reduce the number of trips made to the station. The technology is already applied in 4G LTE, and with the emergence of 5G, the first responders will be able to use large video and audio files with exceptional resolution power in their reports (Asif 2018). Furthermore, data in the reports will be accessible for any queries both from the public and other professional organs. Notwell (2018) noted that descriptions, behaviors, and trends of people will be easily cross-referenced and utilized by emergency responders.

It is anticipated that public safety communications will significantly benefit and will find use cases of the fifth-generation network including its ultra-high frequency capabilities as outlined in Figure 3.10. Table 3.2 also depicts the performance targets of the fifth-generation network.

3.7 Applying 4G and 5G Networks in Public Safety

The fourth- and fifth-generation wireless communication networks can provide public safety community capabilities to enhance their responsiveness to emergencies. The main reasons why public safety needs to adopt fourth- and fifth-generation technologies is for their push-to-talk voice as well as mission-critical communications. The sector is increasingly voice-centered from the public safety answering point (PSAP), which is usually an emergency massage-passing center that answers call from 911 (Bishop 2017). In the recent past, the 3GPP developed functionalities and specifications needed for future public safety communication. For instance, system architecture 6 (SA6) involves working group standards that assess public safety communication and approves 3GPP plans to create policies that support their requirements. Some of the desired 4G and 5G

Table 3.2 Performance targets of the fifth-generation network.

Capability	Explanation	Target	Usage
Latency	Radio network and contributions to the packet travel speed	1 ms	URLLC
Data rate (peak)	The maximum achievable data speed	20 Gbps	eMBB
Energy efficiency	The data sent or received per unit energy usage	Similar to 4G	eMBB
Mobility	QoS requirements and the maximum handoff speed	500 km h^{-1}	URLLC/eMBB
User data rate	The achievable data across areas covered	1 Gbps	eMBB
Connection density	The number of devices per unit area	10^6 km^{-2}	MMTC
Area trafficability	The amount of traffic in the covered area	1000 (Mbps) m^{-2}	eMBB
Spectrum efficiency	Throughput/wireless bandwidth and per network cell	3-4x 4G	eMBB

capabilities in public safety include two-factor authentication, instant data access, the capability to operate in harsh conditions, body camera, mission-critical push-to-talk, loudspeaker phones, and haptic feedback (involving sense-to-touch interfaces) (Bishop 2017). Meanwhile, an initiative by the European Union has asked 3GPP to reflect on 74 descriptions (use cases) regarding how the technology will be deployed.

Some of the cornerstone cases of 4G and 5G include enhanced mobile broadband (eMBB) involving video downloads that can support augmented reality, 4K video and virtual reality. Another important development includes the massive machine type communications (MMTCs) that can be utilized to communications with numerous devices (Wei et al. 2017). Furthermore, the technologies deliver ultra-reliable as well as low-latency communication (LLTC). Ultra-reliability attributes are considered to be essential in voice-to-voice (V2V) communications, particularly in emergencies (Bishop 2017). 4G LTE has a latency of 10 ms with 5G intending to reduce the latency to 1 ms, as depicted in Figure 3.11.

In 4G and 5G networks, more functionalities are moving to the network edge, and end-to-end latency is significantly reduced to give the public safety outstanding technologies to deploy.

Nonetheless, instead of offering a generic recommendation for the implementation of these technologies, it is very important to address some fundamental queries concerning standardization, public safety solutions, and 3GPP releases (Dahlman et al. 2018).

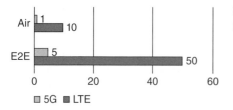

Figure 3.11 Latency of 4G LTE and 5G. Source: Bishop (2017).

3.7.1 The Right Time to Implement 3GPP in Public Safety

It is vital to understand that the currently used technologies in public safety depend on 3GPP standardization. For instance, in Europe, over 1.5 million public safety actors use mainstream technologies including consumer networks (Rodríguez and John Wiley & Sons 2015). Nonetheless, it is not about the user but when the technologies will be applied in 3GPP. 3GPP has relentlessly developed various standards, as outlined in the subsequent section. As a result, it is the obligation of the users to determine the urgency of utilizing 3GPP standards since 3GPP have already done enough with the introduction of standalone and NSA specification in 2018.

3.7.1.1 3GPP

The 3rd Generation Partnership Project (3GPP) involves cooperation between telecommunications associations for developing standards called Organizational Partners (Haneda et al. 2016). Their initial objective was to develop universally applicable third-generation mobile phone specifications guided by the Global System for Mobile Communications (GSM) in the realms of International Mobile Telecommunications-2000 under ITU. However, their scope was later extended to encompass development and maintenance of GSM and associated 2G and 2.5G standards (Enhanced Data rates for GSM Evolution [EDGE] and the General Packet Radio Service [GPRS]) (Shen et al. 2012). Also, 3GPP is charged with the responsibility of 3G standards (such as High-Speed Packet Access [HSPA]), 4G and LTE standards as well as LTE-Advanced and LTE-Advanced Pro. Furthermore, the groups are developing standards for next generation and 5G as well as IP multimedia subsystems (IMSs). 3GPP standardization involves a radio access network, a core network, terminals, as well as services and systems (Vinel 2012). The organizational partners originate from North America, Asia, and Europe and intend to establish the general policies and strategies for 3GPP and undertake the approval and maintenance of 3GPP (Haneda et al. 2016). The standards of 3GPP are included in Table 3.3.

Every release integrates hundreds of standards, and most of the standards have undergone several revisions in the subsequent standards (Vinel 2012). The current 3GPP outlined in Table 3.3 involves the latest revisions. During their work, 3GPP has collaborated with various groups that represent the public safety communication discipline (Shen et al. 2012). Such collaboration is intended to ensure that public safety meets requirements.

Beginning from Release 12, LTE standards offer features concerning the proximity of services as well as group call enablers that 5G and LTE have utilized as an integral part of broadband in PSNs (Furht and Ahson 2016). This has provided public safety community economic benefits including the increasing deployment of commercial off-the-shelf (COTS) technology and technical enhancements such as improved data rates as well as multimedia communication. The commercial cellular operators have gained access to novel capabilities that allow new kind of business and consumer services. Additionally, such development plays a vital role in providing public safety communications through LTE and 5G networks (Furht and Ahson 2016).

Following the freeze of 5G NSA specifications in 2018, integrity and interoperability are being achieved for 5G new radio (NR) specifications. This was a major step within the discipline of a next-generation cellular network (Ahmadi 2018). In 2017 December,

Table 3.3 3GPP standards.

Version	Time	Features
Phase 1	1992	GSM attributes
Phase 2	1995	GSM attributes (EFR Codec)
96	1997	GSM features and 14.4 kbps data rate
97	1998	GSM features for GPRS
98	1999	GSM features EDGE, AMR, and GPRS for PCS1900
Release 4	2001	Release 2000 adding features to the all-IP core network
Release 5	2002	The introduction of HSDPA and IMS
Release 6	2004	Introduction of integrated operation for wireless LAN networks as well as adding MBMS, HSUPA enhancements to IMS like push-to-talk over cellular
Release 7	2007	Decreasing latency, enhancing QoS, and real-time apps like VoIP.
Release 8	2008	Introduction of the first LTE, new OFDMA, MIMO, and FDE based on radio interfaces
Release 9	2009	WiMAX and LTE/UMTS interoperability
Release 10	2011	LTE advanced to fulfill IMT advance 4G requirements
Release 11	2012	Advanced IP interconnection services
Release 12	2015	Enhancements of small cells
Release 13	2016	LTE in unlicensed and LTE enhancements for machine-type communications
Release 14	2017	Mission critical data over LTE, flexible mobile service steering (FMSS), location services (LCS) mission critical video over LTE
Release 15	2018	First new radio (NR) release. Support for the 5G vehicle-to-x service, Future railway network mobile communications and IP multimedia core network subsystem (IMS).

Source: Haneda et al. (2016).

3GPP announced the completion of the 5G standard. The difference with the recent development is that the 2017 specification was intended for NSA 5G NR that was being built on top of the existing LTE networks. The new specification on a standalone version for 5G will enable new usage in areas that do not have an existing infrastructure (Ahmadi 2018). The completion of standalone specifications is believed to complement NSA specifications giving 5G NR the capability of being deployed independently.

Furthermore, it offers a new end-to-end network architecture that makes 5G a facilitator as well as an accelerator for communications and intelligent information technology progression processes. New business models will utilize interconnected things for both industrial and mobile operators (Haneda et al. 2016). Therefore, since the very significant two halves of 5G have been completed, less work is required to finalize its deployment. Some of the 5G developments underway include chips, antennas, hardware, phones, and modems.

The standard compliant 4G push-to-talk (PTT) voice services might start immediately after the equipment providers have tested the public safety devices to ensure they are fully operational.

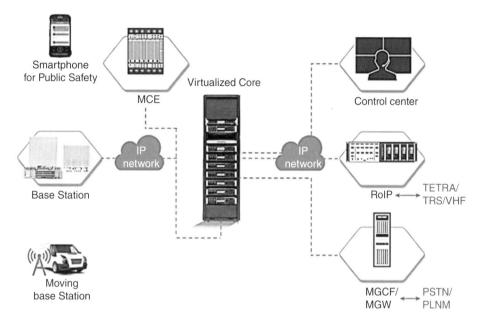

Figure 3.12 Deployment of 4G LTE in unifying networks in public safety communication. Source: Samsung (2018).

3.7.2 4G LTE as a Basis for Public Safety Communication Implementation

4G LTE is a good basis for the implementation of public safety communication. The technology delivers high transference of data instantaneously. Thus, it delivers high-quality images as well as videos that can be sent via many applications like live streaming during an emergency (Samsung 2018). LTE can unify the existing PSNs that are different, such as Tetra, VHF, and UHF. In Figure 3.12, we examine how 4G LTE unifies several networks.

Moreover, LTE has been found to rigorous satisfy the PSN requirements of QoS, resilience, and security.

3.7.3 Implementation of 5G in Public Safety

While 5G offers superior services in public safety, 4G LTE is already being considered as the starting point for many actors. 5G implementation in public safety will encompass all functionalities of 4G and 4G LTE. Stakeholders should not wait until the completion of 5G standards and commercialization (Hu 2016). They should start by tapping into the massive benefits of 4G and LTE to avoid the long wait. As discussed earlier, 5G provides superior capabilities to 4G and LTE networks, but public safety should not undermine the benefits of LTE, whose standards are universally accepted (Dahlman et al. 2018).

3.8 Conclusion

In daily life and during emergencies, rigorous and update communications systems are considered mission-critical in saving lives, preserving communities, and safeguarding property. During man-made or natural disasters, an effective communication system needs to be swift and deliver reliable communications. To achieve the role of public safety actors as well as first responders who rely on robust communications, it is important to use ultra-modern technologies. The chapter has explored the evolution of various communication systems and networks. It is obvious that the discipline has undergone significant advancements intended to strengthen national preparedness and enhance emergency communication.

Nonetheless, they are not deploying the state-of-the-art networks and communications systems offered by 4G and 5G. It is important for the conventional and modern technologies to work concurrently until all functionalities for the 5G and 4G LTE are developed. Furthermore, it is important that the stakeholders in public safety come together to ensure various requirements (funding, governance, resiliency, research and development, guaranteed access, roaming, QoS, standards definitions, reliability, and talk around/direct/simplex mode) are met that facilitate the deployment of future technologies.

References

Abdelhadi, A. and Clancy, C. (2015). An optimal resource allocation with joint carrier aggregation in 4G-LTE. In: *Computing, Networking and Communications (ICNC), 2015 International Conference on*, 138–142. IEEE.

Ahmadi, S. (2018). *5G: Architecture, Technology, Implementation, and Operation of 3GPP New Radio Standards*. S.l.: Elsevier Academic Press.

Asif, S.Z. (2018). *5G Mobile Communications: Concepts and Technologies*. Boca Raton, FL: CRC Press/Taylor & Francis Group.

Bader, F, Martinod, I.., Baldini, G. et al. (2017). Future evolution of public safety communications in the 5G Era. *Transactions on Emerging Telecommunications Technologies* 28 (3): e3101.

Baldini, G., Ferrús Ferré, R. A., Sallent Roig, J. O., Hirst, P., Delmas, S., & Pisz, R. (2012). The evolution of public safety communications in Europe: the results from the FP7 HELP project. ETSI Reconfigurable Radio Systems Workshop: 12 December 2012: Sophia Antipolis, France (pp. 1–10). https://pdfs.semanticscholar.org/f2a6/9d3bd388cb77b9e348688625d4d74a3268cd.pdf

Baldini, G., Karanasios, S., Allen, D., and Vergari, F. (2014). Survey of wireless communication technologies for public safety. *IEEE Communications Surveys & Tutorials* 16 (2): 619–641.

Bishop, D. (2017). *Why Public Safety Communications Needs 5G*. agl Media Group http://www.aglmediagroup.com/why-public-safety-communications-needs-5g

Blind, K. (2016). The impact of standardization and standards on innovation. In: *Handbook of Innovation Policy Impact*, 423. Edward Elgar Publishing.

Bonneville, H., Brunel, L., and Mottier, D. (2016). Standardization roadmap for next train radio telecommunication systems. In: *International Workshop on Communication Technologies for Vehicles*, 51–61. Cham: Springer.

Carla, L., Fantacci, R., Gei, F. et al. (2016). LTE enhancements for public safety and security communications to support group multimedia communications. *IEEE Network* 30 (1): 80–85.

Chen, L. and Mitchell, C. (eds.) (2014). *Security Standardisation Research: First International Conference, SSR 2014, London, UK, December 16–17, 2014. Proceedings*, vol. 8893. Springer.

Chow, A. T., Robert, R. M. I., Murray, J. F., & Rice, C. W. (2012). *U.S. Patent No. 8,179,820*. Washington, DC: U.S. Patent and Trademark Office.

Cox, C. (2012). *An Introduction to LTE: LTE, LTE-Advanced, SAE and 4G Mobile Communications*. Wiley.

Dahlman, E., Parkvall, S., and Skold, J. (2013). *4G: LTE/LTE-Advanced for Mobile Broadband*. Academic Press.

Dahlman, E., Parkvall, S., and Skold, J. (2018). *5G NR: The Next Generation Wireless Access Technology*. Academic Press.

Doumi, T., Dolan, M.F., Tatesh, S. et al. (2013). LTE for public safety networks. *IEEE Communications Magazine* 51 (2): 106–112.

Fantacci, R., Gei, F., Marabissi, D., and Micciullo, L. (2016). Public safety networks evolution toward broadband: sharing infrastructures and spectrum with commercial systems. *IEEE Communications Magazine* 54 (4): 24–30.

Ferrus, R. and Sallent, O. (2015). *Mobile Broadband Communications for Public Safety: The Road Ahead through LTE Technology*. Hoboken, NJ: Wiley.

Ferrus, R., Pisz, R., Sallent, O., and Baldini, G. (2013a). Public safety mobile broadband: a techno-economic perspective. *IEEE Vehicular Technology Magazine* 8 (2): 28–36.

Ferrus, R., Sallent, O., Baldini, G., and Goratti, L. (2013b). LTE: the technology driver for future public safety communications. *IEEE Communications Magazine* 51 (10): 154–161.

Fodor, G., Parkvall, S., Sorrentino, S. et al. (2014). Device-to-device communications for national security and public safety. *IEEE Access* 2: 1510–1520.

Furht, B. and Ahson, S.A. (eds.) (2016). *Long Term Evolution: 3GPP LTE Radio and Cellular Technology*. CRC Press.

Gomez, K., Goratti, L., Rasheed, T., & Reynaud, L. (2014). Enabling disaster resilient 4G mobile communication networks *arXiv:1406.0928*.

Haneda, K., Tian, L., Asplund, H. et al. (2016). Indoor 5G 3GPP-like channel models for office and shopping mall environments. In: *Communications Workshops (ICC), 2016 IEEE International Conference on*, 694–699. IEEE.

Hu, F. (ed.) (2016). *Opportunities in 5G Networks: A Research and Development Perspective*. CRC Press.

Huang, J., Qian, F., Gerber, A. et al. (2012, June). A close examination of the performance and power characteristics of 4G LTE networks. In: *Proceedings of the 10th International Conference on Mobile Systems, Applications, and Services*, 225–238. ACM.

Iapichino, G., Câmara, D., Bonnet, C., and Filali, F. (2014). Public safety networks. In: *Crisis Management: Concepts, Methodologies, Tools, and Applications*, 376–393. IGI Global.

Khan, M.A., Saeed, N., Ahmad, A.W., and Lee, C. (2017). Location awareness in 5G networks using RSS measurements for public safety applications. *IEEE Access* 5: 21753–21762.

Kumbhar, A. and Güvenç, I. (2015). A comparative study of land mobile radio and LTE-based public safety communications. In: *SoutheastCon 2015*, 1–8. IEEE.

Kumbhar, A., Koohifar, F., Güvenç, I., and Mueller, B. (2017). A survey on the legacy and emerging technologies for public safety communications. *IEEE Communications Surveys & Tutorials* 19 (1): 97–124.

Lee, J., Kim, Y., Kwak, Y. et al. (2016). LTE-advanced in 3GPP Rel-13/14: an evolution toward 5G. *IEEE Communications Magazine* 54 (3): 36–42.

Li, S.Q., Chen, Z., Yu, Q.Y. et al. (2013). Toward future public safety communications: the broadband wireless trunking project in China. *IEEE Vehicular Technology Magazine* 8 (2): 55–63.

Li, X., Guo, D., Yin, H., and Wei, G. (2015). Drone-assisted public safety wireless broadband network. In: *Wireless Communications and Networking Conference Workshops (WCNCW), 2015 IEEE*, 323–328. IEEE.

Lien, S.Y., Chien, C.C., Tseng, F.M., and Ho, T.C. (2016). 3GPP device-to-device communications for beyond 4G cellular networks. *IEEE Communications Magazine* 54 (3): 29–35.

Lin, X., Andrews, J., Ghosh, A., and Ratasuk, R. (2014). An overview of 3GPP device-to-device proximity services. *IEEE Communications Magazine* 52 (4): 40–48.

Lundgren, R.E. and McMakin, A.H. (2018). *Risk Communication: A Handbook for Communicating Environmental, Safety, and Health Risks*. Wiley.

Luo, F.-L. and Zhang, C. (2016). *Signal Processing for 5G: Algorithms and Implementations*. Chichester, West Sussex, UK: Wiley/IEEE Press.

Management Association, Information Resources (2013). *Crisis Management*. Hershey: IGI Global. Print.

Martinez, D. M., & Vaughan, J. (2012). *U.S. Patent No. 8,145,262*. Washington, DC: U.S. Patent and Trademark Office.

Naqvi, S.A.R., Hassan, S.A., Pervaiz, H., and Ni, Q. (2018). Drone-aided communication as a key enabler for 5G and resilient public safety networks. *IEEE Communications Magazine* 56 (1): 36–42.

Notwell, L. (2018, July 25). *3 Ways First Responders Can Take Advantage of 5G and IoT*. State Tech Magazine: https://statetechmagazine.com/article/2018/07/3-ways-first-responders-can-take-advantage-5g-and-iot

Ohtsuji, T., Muraoka, K., Aminaka, H. et al. (2018). Relay selection scheme based on path throughput for device-to-device communication in public safety LTE. *IEICE Transactions on Communications* 101 (5): 1319–1327.

Paulson, A. and Schwengler, T. (2013). A review of public safety communications, from LMR to voice over LTE (VoLT E). In: *Personal Indoor and Mobile Radio Communications (PIMRC), 2013 IEEE 24th International Symposium on*, 3513–3517. IEEE.

Rizzo, C. (2011). ETSI security standardization. In: *The Proceedings of 2011 9th International Conference on Reliability, Maintainability and Safety*, 633–638. IEEE.

Rodríguez, J. and John Wiley & Sons (2015). *Fundamentals of 5G Mobile Networks*. Chichester: Wiley.

Rouse, M., Gerwig, K., & Haughn, M. (2018). *5G*: Newton, MA Search Networking https://searchnetworking.techtarget.com/definition/5G

Samsung. (2018). *Public Safety LTE (PS-LTE)*. Seoul, South Korea Samsung: https://www.samsung.com/global/business/networks/solutions/public-safety-lte

Shajaiah, H., Abdel-Hadi, A., and Clancy, C. (2014). Spectrum sharing between public safety and commercial users in 4G-LTE. In: *Computing, Networking and Communications (ICNC), 2014 International Conference on*, 674–679. IEEE.

Shen, Z., Papasakellariou, A., Montojo, J. et al. (2012). Overview of 3GPP LTE-advanced carrier aggregation for 4G wireless communications. *IEEE Communications Magazine* 50 (2): 122–130.

Su, X., Castiglione, A., Esposito, C., and Choi, C. (2018). Power domain NOMA to support group communication in public safety networks. *Future Generation Computer Systems* 84: 228–238.

Sun, S., Gao, Q., Chen, W. et al. (2015). Recent progress of long-term evolution device-to-device in third-generation partnership project standardization. *IET Communications* 9 (3): 412–420.

TCCA. (2017). *4G and 5G White Paper*. LTE Technology: Heemskerk, The Netherlands http://www.criticalcommunicationsreview.com/lte/news/88584/tcca-publishes-whitepaper-4g-and-5g-for-public-safety

Thomas, J., & Squirini, A. (2018). Measuring Systems Interoperability: A Focal Point for Standardized Assessment of Regional Disaster Resilience. http://www.spinglobal.co/wp-content/uploads/2018/04/Measuring-Systems-Interoperability.pdf

Vinel, A. (2012). 3GPP LTE versus IEEE 802.11 p/WAVE: which technology is able to support cooperative vehicular safety applications? *IEEE Wireless Communications Letters* 1 (2): 125–128.

Wei, X., Shen, X., and Zheng, K. (2017). *5G mobile communications*. Cham: Springer.

Williams, D. (2017). Time for a 5G technology assessment and road map? [President's column]. *IEEE Microwave Magazine* 18 (6): 10–14.

4

Keys to Building a Reliable Public Safety Communications Network

4.1 Introduction

According to Future Mobile Broadband PPDR Communications Systems (2015), a reliable public safety communications network has guidelines for optimal operations. One of the guidelines refers to the duty to serve requirement. Duty to serve implies that the public safety communications network must be available for, and is expected to be at, the right position to work when scheduled regardless of events that are going on. It needs to work even during extreme events like:

- Flooding
- Earthquakes
- Hurricanes
- Man-made disasters like terrorist attacks
- Wildfires.

These systems must be in working shape in order to facilitate the flawless and efficient work of emergency response workers like firefighters, paramedics, etc. Duty to serve extends to enabling seamless communication among the respondents by creating, a safe, secure, and efficient system, which can create priorities for communication between the respondents. The modernization of networks, applications, broadband internet, and social media has become an opportunity and challenge for stakeholders in public safety. Thus, policies have been developed to engage the "whole community" in national preparedness activities. The department of Homeland Security (DHS) in the USA has worked with municipalities to update the National Emergency Communications Plan. In the plan, voice, video, and data will be used to communicate and share information in all instances. The success of the plan is contingent upon the emergency communications community.

4.2 Supporting the Law Enforcement Elements of Communication

The system must have the capacity to support the following needs and requirements of communication by the first responders and law enforcement agencies:

- Taking of calls and making dispatches
- Facilitation of the emergency resource management and of the law enforcement unit

- Ensuring officer safety
- Creation of a system for handling reports
- Facilitation of special situation implementation of law enforcement agencies (vehicle or foot pursuit).

These systems must also help in the facilitation of fire services management. Here, the core handling capabilities include the following:

- Dispatches and call taking
- The resource controlling the local and national fire service apparatus
- The facilitation of fire alarm dispatching
- Conducting and coordinating the state fire incident command systems
- Facilitation of quick communication concerning hazardous materials.

public safety communications networks must also have the capacity to enhance the emergency medical services (EMS) communications elements such as:

- Fast and secure call taking and dispatch of the (EMS) personnel
- Efficient management of the allocation of EMS resources
- Enhanced management of EMS units
- Integration of specialized training in taking caller questions
- Facilitation of quick medical instructions on the telephone.

4.3   Components of Efficient Public Safety Communication Networks

The five critical components which are important and need to be considered when planning and implementing a reliable solution for all public safety communications technologies are:

1. Facilitation of communication between the emergency responders and law enforcement agencies with the public
2. Communicate safely, faster, and efficiently between members of the same agency
3. Communication through an interoperable format between the public safety agencies
4. Safe and efficient communication between public safety agencies
5. Secure, fast, and efficient communication between the people and public safety agencies and other relevant support services.

For the system to be efficient, it must be able to provide an aspect called mutual aid. Mutual aid is a process that facilitates the supply of supplemental personnel and equipment to the agencies that might be facing being overwhelmed when they are responding to an emergency (Bazzo et al. 2017).

Mutual aid is an important safe and secure protocol that takes into account the supplies of supplementary personnel and ensures that at all times the personnel responding have support systems that not only keep them safe but also keeps citizens safe, and guarantees good quality services to citizens.

4.4   Networks Go Commercial

With mobile broadband delivered by commercial network operators everywhere, consumers have for a long time been accessing powerful smartphone applications in a big way. Seemingly, there is an application or several applications for anything that people want to do.

Public safety authorities such as the fire services and police officers, and other emergency responders are digital natives. These people being digital natives have their sights on several potential applications, which will make their work much simpler and more efficient (Bazzo et al. 2017).

They are seeking applications that will not only make their work easier but also make their work safer to do with a lot of precision. They therefore want to augment their narrowband services with broadband services in the incumbent RED band zone to transform how they work.

These requirements raise an important question concerning how these services will come into existence. Traditionally, public safety organizations have used their own dedicated professional mobile radio networks to execute their public safety duties.

However, in today's mobile broadband driven mobile network, and operators with increased expertise and infrastructure, it is becoming more demanding for the mobile network operators to come up with ways of meeting the stringent requirements of public safety. In their assessments, the mobile network operators seek to find out if these ventures are profitable to them, thus necessitating the need for public safety to go commercial (Bazzo et al. 2017).

4.5 Viable Business Prospects

There are several models in existence in which the government and operators are able to work together cooperatively in order to produce network services that can meet the increasing public safety communications needs, while at the same time giving a solid business case that the mobile network operators can use to go commercial.

With the reality of several funding possibilities, which are available and can provide the significant investments required for the mobile network operators to provide Long Term Evolution (LTE), which is critical for public safety missions, there is a strong case for going public with the LTE service. These viabilities create plausible revenue models, which bring different options for commercial pricing on which governments and public safety authorities can agree (End-User Devices Connected to Critical Communications Systems 2018).

In order to satisfy the mission-critical requirements, there are three key areas that require investment. These areas also present viable cases for mobile network operators to go commercial with public safety networks. The three key areas are as follows.

4.5.1 The Core Network

- The core network comprises dedicated network elements, which are deployable to government facilities and are secure.
- The dedicated networks could also be deployed in the premises of the mobile network operators.
- Virtualized core networks are possible to use in order to flexibly maintain and deploy dedicated gateways with the aim of maintaining high-security levels.

4.5.2 The Radio Network

- The radio network guarantees the availability of mission-critical network coverage of LTE and the viability of extending the capabilities by the installation of new sites while hardening the existing sites.

- Putting in place rapidly deployable LTE systems that are usable during extreme conditions.
- The radio networks are shareable by commercial users and public safety authorities controlled by policy mechanisms and access policy mechanisms.
- The development of applications that are in nature dedicated to public safety.
- These applications include app services like push-to-video and push-to-talk, which aim to enhance quality communication by the first responders and the police

4.6 The Industry Supports the Involvement of the Mobile Network Operators

Public safety networks going commercial is also because of some movement in industrial market dynamics. One of these dynamics is that the opportunities for the involvement of the public safety networks come from the underscored escalation in agreement within the industry concerning the important roles that the mobile network operators play in public safety and emergency response to improve efficiency (Moore and Library of Congress 2016).

The wireless bodies in industries like the Global Suppliers Association, and the TCCA, or the TETRA Critical Communications Association, are running rigorous campaign efforts that are promoting the partnerships between the mobile network operators and the safety organizations, which are responsible for the safety of the public.

This level of involvement is evidence of the increasing involvement and encouragement of the public safety networks going commercial. A good example is a successful collaboration between the United Kingdom Government and the organization EE.

These two entities are working collaboratively through public–private commercial cooperation in order to create efficient and effective emergency services networks (ESN), which EE will run over its public radio network. This is an indication that there is increasing encouragement from the private industry for government involvement (Moore and Library of Congress 2016).

This collaboration ensures that the core network dedicated to government users processes through the public safety user protocols provided by the private players in the industry. In the industry, Nokia Company is working with EE with a view to developing ESN. AT&T and FirstNet in the United States are other examples of private industries working to provide safe communication capabilities for the public good and efficiency in essential services response capabilities (Moore and Library of Congress 2016).

Globally, several administrations are seeking to leverage the advantages of the increasing mobile broadband technologies in order to improve the capabilities of the emergency service communities and the police communities so as to provide premium and quick services to citizens across the globe.

In connection with these efforts, these administrations are employing various models that are already working in the mobile network operator infrastructural systems. Governments, therefore, are getting into strategic partnerships with these mobile infrastructural companies to upscale their networks and the quality of services they offer to security officers and first responders. These governments are providing applications and network services that will be helpful in meeting the increasing demand to give first responders better equipment, which increases the situational awareness that enhances the responsiveness of these agencies in providing services to the public (The United States 2016).

Some of the prominent examples of private companies that are helping governments improve their data, security, and communication efficiency are Nokia, Critcomm Insights, and AT&T, which are actively engaging in analysis and provision of mission-critical communication capabilities to governments.

These entities are working on providing their services based on three broad areas, namely:

- Nokia is concentrating on several aspects of business considerations of interest, both to public safety agencies and the mobile network operators, regarding the provision of a secure means of communication between critical government safety and responder agencies.
- Certicom Insight is concentrating on analyzing how it can cover the important mission-critical technology, data communications, the global trends in communications security, and the increasing role the mobile network operators play in the creation of a safe and secure means of communication to enhance the efficiency capacities of first responders and security apparatus.
- FirstNet is an important private stakeholder in the government responder responsiveness in service provision. FirstNet is the leading public–private partnership that has the responsibility of building and delivering America's nationwide platform dedicated to communications of public safety matters.
- AT&T, on the other hand, will play the role of providing a view into serving the public safely, geared to meeting the rigorous requirements of the standards for efficiency and communication.

In working with the government mobile network operators do not only win the fees collected from the government for the operational services they provide. The networks, which meet the mission-critical needs of the users, also develop the highest reliability and capacity for efficient provision of reliable services to its client base.

When a company achieves this stature, it is able to minimize churn. Companies that act as important links for emergency services in country jurisdictions raise their prospects for other highly critical and beneficial businesses as well as enhancement of their reputations. Mobile network operators and public safety organizations may have good working relationships, which could extend their work dealings to the professional radio and broadband sectors.

4.7 Policies for Public Safety Use of Commercial Wireless Networks

Naturally, emergency safety first responders obtain a great deal of advantage from wireless broadband functionality because the wireless broadband capabilities can bring the services that the first responders have so far never enjoyed (The United States 2016).

Through the provision of roaming services to safety responders, access to commercial networks can increase, on a priority basis, service reliability, and geographic coverage to first responders. These users can, therefore, enjoy services that they have never had before, or that, which would be possible within the networks, are dedicated only to public safety users. The advantages of roaming are subject in part to the way in which the priority access is structured.

4.8 Public Safety Networks Coverage: Availability and Reliability Even During Outages

Public safety is paramount at all times. It becomes even more critical in times of disasters such as tsunamis, earthquakes, flooding, mass casualty accidents, and terrorist attacks. In recent decades, there has been an escalation in the number of natural disasters like earthquakes and hurricanes. During these events, the telecommunication networks are often not spared the serious damage caused by these events (Saunders 2015).

This damage sometimes hamper the capabilities of the first responders to offer rescue services because of poor communication due to extensive damage. These incidents, therefore, form an interesting case study for the hardening of the telecommunication capabilities for the first responders to make them able to successfully address accidents and incidences.

The Federal Communications Commission (FCC), charged with making communications possible in the country, conducted a study post-2006 Hurricane Katrina in order to study the impact that the hurricane had on the infrastructure of telecommunications, including the public safety networks in the United States (The United States 2016).

The study came up with important lessons, which also provided critical guidelines for enhancing the emergency communications for both the core network and radio network access. As seen recently, there were lessons learned that provided a case for mitigating against the effects of these outages.

4.9 FirstNet Interoperability

Year after year, interoperability has been ranked increasingly as one of the most crucial communication issues when it comes to public safety. It influences the ability of agencies to respond efficiently to protection matters when it comes to personal safety as well as responding to catastrophic events during joint response interventions (Saunders 2015).

Entities responsible for public safety all over the country continue to fight this issue to fluctuating levels of success. Now, with the establishment of public safety broadband communications and FirstNet on the horizon, a set of new complications will be added to the already existing complications that have caused trouble to the agency for years.

The two new systems have the ability to ensure the provision of advanced capabilities, which will help to provide new tools for addressing interoperability solutions meant for public safety throughout the country that agencies have needed for quite some time (Saunders 2015).

However, the two networks will bring about technical challenges that have never been witnessed in the public safety field before, despite providing commendable benefits. The main theme for agencies while adopting this advanced technology will be the utilization of a layered approach to challenges associated with interoperability.

As the strides in advancing technology take place there is an increased possibility that benefits will accrue from it but there is also a possibility that the new technology will come with complexities adding up to challenges associated with interoperability. Each and every year there is a high profile of events that need high communication interoperability levels including mass shootings, natural disasters, and terrorist attacks. The events mentioned above need close collaboration and support from different agencies. They also require a significant and immediate response from different agencies (Future Mobile Broadband PPDR Communications Systems 2015).

Over the last few decades, a sense of a high degree of concentration and focus, coupled with intervention efforts, has been made by the local, state, and federal agencies in collaboration with partners from the industry to plan and develop standards that will enhance capabilities for interoperability. Project 25 (P25) acted as a major step forward toward common standards that all agencies from the local, state, and federal government could use.

Standards for the P25 suite, along with advancements in radio technology, have increased the level of prevalence for multi-band radios to permit a single radio to pass information across multiple bands regardless of whether it is UHF, VHF, 700 or 800 MHz. This has allowed users to overcome the obstacles that existed a while ago between systems that used differing bands, which earlier relied on the utilization of inefficient methods such as bridging systems or multiple radios with shared distance coverage agencies (Future Mobile Broadband PPDR Communications Systems 2015).

With the establishment of FirstNet and the upcoming National Public Safety Broadband Network (NPSBN), yet another advanced technology to supplement to the mix meant for utilization of data applications primarily, there will ultimately be a requirement to interface to the existing land mobile radio (LMR) networks to establish an all-inclusive communications solution.

Future capabilities to ensure delivery of voice-over-LTE will bring important capabilities for the agencies since it will provide a unified voice as well as a data network to one subscriber gadget or device. Strategies to ensure proper integration of these networks alongside already existing communications systems in a complementary manner, and in some instances migrate the already existing LMR, will be very important and involve extra levels of technical complication, adding up to an already existing thorny issue (Desourdis et al. 2015).

4.10 Solutions for Enhancing Availability and Reliability Even During Outages

- Increasing the radio access network (RAN) resiliency
- Hardening of the cellular sites as viewed as effective.

4.11 National Public Safety Broadband Network (NPSBN)

Through the Middle-Class Tax Relief and Job Creation Act created in 2012, an independent authority was established through the NTIA, or the National Telecommunications and Information Administration: the First Responder Network Authority or FirstNet. FirstNet has the responsibility of not only managing the 700 MHz spectrum, but also for the designing, planning, and building of the NPSBN (Bazzo et al. 2017).

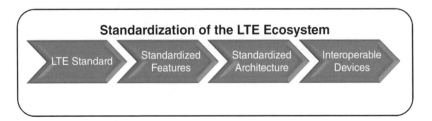

Figure 4.1 FirstNet standardization of the LTE ecosystem. Source: http://commdex.com/wp-content/uploads/2018/05/Commdex-Whitepaper-Interoperability-Considerations-Between-P25-and-LTE-Networks.

This takes care of the needs of local, tribal, state, and federal public safety agencies with its efforts geared toward addressing as many different needs and concerns of the users and the community as possible.

After the conclusion of the process of consulting with the representatives in every region and every state, the network was deployed in regions throughout the country. The states that approved the plan at the end of the implementation had RAN, LTE, and FCC approvals for having the system in place.

Through adoption of the LTE as a plan, the NPSBN in the United States has been using the existing commercial standards as well as adopting these standards for maximization of public safety. This system brought economies of scale in the implementation of the purchase of the network security equipment as well as the creation of amalgamated networks to create extended capabilities. Figure 4.1 shows an example of LTE and FirstNet standardization.

Taking these combined capabilities and adopting them in public safety has been a primary concern for the community of public user safety. FirstNet, therefore, works hand in hand with the communities that require these security systems to take to account of the needs of the various jurisdictions and to adopt them in the FirstNet design that promotes efficient use and secure systems for these communities of users (End-User Devices Connected to Critical Communications Systems 2018).

4.12   Important Objectives of NPSBN

a) One of the key objectives of NPSBN is the provision of a single and interoperable broadband data network that will provide feasible solutions to safety responders at all government levels in the country.
b) Provision of advanced capabilities through a secure and hardened infrastructure.
c) Ensuring that during catastrophic events like terrorist activity, there is a super-fast facilitation of video exchange between law enforcers through aerial units like closed-circuit televisions, drones, and cameras for the apprehension of suspects and mitigation of further damage.
d) Providing the means of managing and disseminating the information that is in the interest of the public with the safety agencies at all levels of operations.

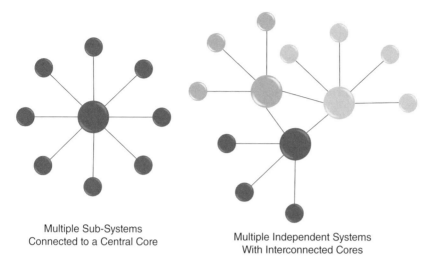

Multiple Sub-Systems
Connected to a Central Core

Multiple Independent Systems
With Interconnected Cores

Figure 4.2 Future of FirstNet interoperability capability. Source: http://commdex.com/wp-content/
uploads/2018/05/Commdex-Whitepaper-Interoperability-Considerations-Between-P25-and-LTE-
Networks.pdf.

4.13 The Future of FirstNet: Connecting Networks Together

Interoperability is the core future of FirstNet. In this regard, FirstNet seeks to create a
network of networks, see Figure 4.2. It has been the goal of the different user groups to
create a large-scale interoperable network. However, this ambition faces the following
challenges:

  i. Technological challenges
 ii. Cost challenges
iii. Operational challenges.

These challenges have been facing the agencies for several years. For instance, there
is a lack of a clear single size that fits all the approaches and therefore different user
groups have to resort to a combination of several of them, or different groups have to
take multiple approaches in order to achieve several connected networks within specific
areas. The implication of this for the future is that FirstNet will develop a plan for building the network and addressing the questions of distributed versus centralized network
architectures (Moore and Library of Congress 2016).

The centralized approach takes into account the creation of a single network that has
a single network identity, which in effect will be a single service provider to various users
of a wireless broadband to the entire public safety community.

In such a creation, FirstNet may establish a user fee, which will help in the maintenance
and the operationalization of the network at the local level of administrative control.
These costs would consequently be minimal because of the efficiency and the seamless
working of the data and security services provided by FirstNet. Any agreements over
roaming services with carrier networks will be under a national management basis, with
a single agreement, which would ensure the coverage of all the users of NPSBN (Moore
and Library of Congress 2016).

Some of the challenges with this approach include:

a. The local input could be extremely critical and beyond the capability of the network to be able to satisfy the needs of the different agencies across the country.
b. The systems will need to undergo development and deployment to provide an enhanced level of control to the local agencies, which will have the users operating on the NPSBN.
c. Since the systems will be developed and thereafter deployed to enhance a level of control to the local agencies, which have users operating on the NPSBN channel, the infrastructure vendors might find themselves having a limited capacity of incorporating this level of control with their current statures.

Ensuring that the local agencies have the ability to manage local capacities, control users, and detect the faults in the network system, are all important issues that may need to be addressed in order to make FirstNet successfully provide a satisfactory service regime to all the necessary agencies throughout the nation.

Alternatively, there would be a less homogenous approach that would allow for regional or state networks, which are deployedand thereafter interconnected to allow for roaming of subscribers between networks.

This measure would have a reflective effect on a network, where a number of states would have the choice of opting out and then building their own networks that they will connect to the NPSBN. The prevailing waiver holder's networks could come into incorporation in the system following this fashion, therefore preserving the network, which has already been completely done as part of the BTOP grants. This will help in achieving the objective of networking all the systems together in synchrony. This approach, however, has the following challenges (Moore and Library of Congress 2016):

i. It has more core sites than the necessary sites, which are minimally necessary for running the network.
ii. It is capital intensive since it requires additional equipment at the core sites in order to ensure there is efficient roaming between networks.
iii. It has overlapping coverage, which exists between several different networks.
iv. The system has variant user fee structures between the different networks.
v. There is likely to be an inconsistency in the software revisions between the different networks.
vi. Multiple roaming agreements with different carriers would need to be signed.

4.14 High Capacity Information Delivery

Since information and data in motion poses a security threat as they are high level targets, FirstNet offers multiple capabilities. It can handle huge volumes of data at high speed. While moving, FirstNet offers an assurance for the sensitive data it handles in real time. It can monitor the safety of moving voice and video data in real time from one site to another over multiple sites because it has a backup plan and disaster recovery plan from the client's premises to the cloud system and back to the client's premises (Moore and Library of Congress 2016).

FirstNet offers certified encryption of the data that it moves, thereby offering an ideal solution for data in motion security, including time-sensitive video and voice streams, for government and enterprise organizations.

4.15 Qualities that Facilitate Efficient High Capacity Information Handling

4.15.1 FirstNet Has a Trustworthy Security System

- It has certified security systems trusted by several institutions like commercial organizations and telecoms companies
- It uses security based in Certified FIPS 140-2 L3 protocol
- Common criteria certification
- NATO cortication
- UC APL certification.

4.15.2 Concentrated Network Performance

- Near-zero overhead
- Operates on microsecond latency.

4.15.3 Simple and Scalable

- It has a "set and forgets" management system
- The total cost of ownership is low.

4.15.4 High Level of Vulnerability Safeguards

- FirstNet has true end-to-end, above-board encryption
- State-of-the-art customer side key administration.

4.16 FirstNet User Equipment

Legally, the FirstNet network needs to be based on minimum technical requirements, which are supposed to satisfy the commercial standards for LTE services or the wireless communications of high-speed information.

In this regard, FirstNet engages in working and standard processes, and it works closely in collaboration with public safety establishments in order to support the creation of functionalities that meet the public needs served by FirstNet users. Most of the current focus is on the international set of standards that will permit FirstNet to offer MCV or mission critical voice the moment the capabilities become available. These same MCV technologies will thereafter work across all standards-based networks and equipment globally. FirstNet has a broad definition of the LTE network, which are grouped in layers (Moore and Library of Congress 2016).

These layers include:

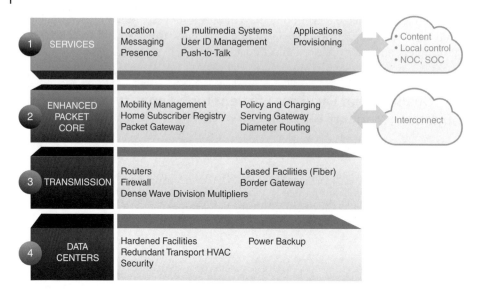

Figure 4.3 Core network. Source: http://firstnetme.gov/process/deployment.html.

i. The transport backhaul
ii. The core network
iii. Public safety devices
iv. Radio access network (RAN).

4.17 Core Network

FirstNet has the responsibility of building an enhanced packet core network, which is a major component that ensures the network has six main functions. Figure 4.3 shows the block layers of the FirstNet core network. These functions are:

i. Processing of data
ii. Reformatting information
iii. Switching data
iv. Storage of data
v. Maintenance of data
vi. Keeping data secure.

The core essentially serves as a huge umbrella, which covers the entire United States, and it has a connection to RAN in every state via the network's backhaul layer.

4.18 Illustration: Layers of the LTE Network

LTE refers to the standard for mobile commercials communication. The key characteristics of LTE technology are the few network elements also known as flat architecture. Figure 4.4 shows network layer of LTE technology.

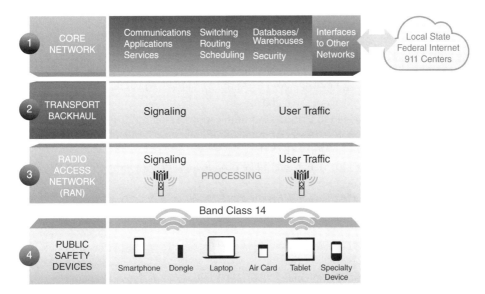

Figure 4.4 LTE network layers. Source: http://firstnetme.gov/process/deployment.html.

4.18.1 Transport Backhaul

Data, voice, and video traffic are carried on the FirstNet network. Backhaul is responsible for providing connections between the core wireless and cell sites broadband network. It also connects the FirstNet to other different networks like 911 centers. These connections are typically made of microwave and fiber optic cables. In order to satisfy the reliability requirements of public safety, backhaul remains redundant wherever possible in order to ensure that the network traffic continues flowing during periods of extreme network stress and demands.

4.18.2 The Radio Access Networks

The RAN part of the network has a radio-base station infrastructure that connects to user devices. The RAN has components like cell towers and mobile hotspots, which are embedded in vehicles, whose function is backhauling to the core network over satellite and other types of wireless infrastructure. The requirements of comprehensive RAN planning optimize coverage, performance, and capacity for a nationwide network.

4.18.3 Public Safety Devices

The devices refer to all the access points of users that will send and receive information over the user network. All devices from tablets, dongles, laptops, and other wide range of special devices are developed for the users of FirstNet. The objective is to come up with rugged devices that can withstand several uses in public safety applications. These devices must also be easy to use, secure, and easy to administer.

References

Bazzo, J.J., Barros, S., Takaki, R. et al. (2017). 4G/LTE Networks for Mission-Critical Operations: A Cognitive Radio Approach. In: *Cognitive Technologies*. Berlin: Springer.

Desourdis, R.I., Dew, R., and O'Brien, M. (2015). *Building the FirstNet Public Safety Broadband Network*. Norwood: Artech House.

End-User Devices Connected to Critical Communications Systems (2018). *Fundamentas of Public Safety Network and Critical Communciations Systems: Technolgies, Deployment, and Mangement*, 165–173. Wiley-IEEE Press.

Ferrús, R. and Sallent, O. (2015). *Future Mobile Broadband PPDR Communications Systems*, 81–124. Wiley.

Moore, L.K. (2016). *The First Responder Network (FirstNet) and Next-Generation Communications for Public Safety: Issues for Congress*. Washington, DC: Library of Congress.

Saunders, M. (2015). *The First Responder Network (FirstNet): Elements, Issues, Progress*. Nova Science Publishers.

Office of NEPA Policy and Compliance (2016). *Nationwide Public Safety Broadband Network Draft Programmatic Environmental Impact Statement for the Central United States*. Washington, DC: Office of NEPA Policy and Compliance.

5

Higher Generation of Mobile Communications and Public Safety

5.1 Introduction

The public safety sector has over the years undergone substantial changes that have affected its operations. In the past, the public safety sector supported its operations by using the narrowband services available and voice-based communication (Department of Homeland Security [DHS] 2011, 2016, p. 1). With the advent of technology, including the growth of networks and the internet, there became a growing demand for the use of high-speed broadband data services (DHS 2011, 2016, p. 1). Over the past few years, the industry has recognized the importance of implementing these features within the public safety sector, which has led to the rise of 4G Long Term Evolution (LTE) and the development of 5G networks. Broadband has become the future of critical communication through its powerful and innovative solution for better protection of public safety. It allows the delivery of real-time information, reliability, and performance of the core goal found in mission-critical technology. As important as the development of 4G LTE is to society, it is a stepping-stone to something greater. Combined with other network elements and the new features as a result of the advancement of technology, LTE has transformed public safety communication. The improvements and features added have paved the way to innovative solutions toward the next generation of technology that can support the increasing need for bandwidth within the frequencies. This chapter will analyze the evolution of the public safety system providing an analysis of the existing public safety network, 4G LTE and its application in public security, its comparison to the 5G network technology, the link between the two, and the spectrum of public safety.

5.2 Review of Existing Public Safety Networks

Currently, in the United States the public safety community bases its operations through the use of the traditional land mobile radio (LMR) systems (DHS 2011, 2016, p. 1). These systems have helped support most of its critical communications in a variety of response environments and operations. While the use of LMR has been effective in public safety, the evolving nature of the world has forced the government to consider the use of other options. As the LMR systems have evolved over the years, they have varied in use, creating an incompatible system within the country (DHS 2011, 2016, p. 3). This has led to challenges in ensuring public safety across different jurisdiction and agency lines. The LMR system is an affordable and an accessible means of communication that helps

Public Safety Networks from LTE to 5G, First Edition. Abdulrahman Yarali.
© 2020 John Wiley & Sons Ltd. Published 2020 by John Wiley & Sons Ltd.

provide voice capabilities through which the public can help in enhancing emergency response (DHS 2011, 2016, p. 1). It has been in use since the 1930s and continues to be the primary means of voice communication that is used for public safety (DHS 2011, 2016, p. 1). As communications have evolved over time, this system has been able to keep up with the pace as it has the ability to be adjusted and improved through the introduction of new features and functionalities. Over the past 50 years, public safety agencies have made huge investments in infrastructure and equipment in LMR. These investments have facilitated the change in the systems from the previously used analog conventional systems. Analog had been used to help provide the basic platforms that were in use for the two-way communication systems. The change occurred with the use of digital trunked systems, which are more advanced in their features, to facilitate higher quality, capacity, longer battery life, and applications that can be used by the emergency responders (DHS 2011, 2016, p. 2). The public safety sector also further invested in the use of P25 standards within the digital systems. These standards have provided a benchmark and the requirements for the equipment in use to help make sure that they can be compatible and support interoperability within its many uses (DHS 2011, 2016, p. 2). The use of the digital LMR public safety has helped in facilitating internet protocol (IP) communication protocols and tools within the infrastructure that help counter any threat of inoperability (DHS 2011, 2016, p. 3). The radio systems in LMR have been used to provide a reliable way through which the personnel in the field can ommunicate with others and those in the control and command centers.

5.2.1 What Are LMR Systems?

LMR systems are wireless communication systems that are terrestrial based and used in major areas of the country by the federal and state government, including in the military (DHS 2011, 2016, p. 1). The functionality of the system normally depends on the use of handset radios by the emergency responders in the field. These devices are usually limited in their range of transmission because they require the responders to be in a certain range of proximity to function. The system also includes mobile radios that are carried in vehicles (DHS 2011, 2016, p. 3). These radios have a larger range of scale because of the presence of the large antennae of the vehicle that can pick up on the signals from a far distance and its power supply helps in ensuring the durability of use (Kumbhar et al. 2016, p. 3). The base radio stations are those that are immobile and are normally found in dispatch centers (Kumbhar et al. 2016, p. 3). These radios normally tend to have the most powerful range of transmission. The systems also have repeaters that are used to help increase the product's range of communication in the whole system by retransmitting the signals that it receives (DHS 2011, 2016, p. 3). Finally, there are the networks that serve the function of connecting the base stations around the area to each other. Many agencies within the country have been standardized to operate their communication in this manner.

In the conventional system, each frequency is designated to different individuals. When a user tries to make a call through these systems and they select a channel, nobody else is able to get access to this channel in the process of the call (Kumbhar et al. 2016, p. 3). When a call is made in LMR, the available channels from the pool are automatically chosen for use. Trunked systems, on the other hand, are controlled by computers and designate different channels to multiple individuals (Kumbhar et al.

2016, p. 3). They have a more complex infrastructure, which provides the capability of allowing the sharing of channels among multiple individuals. This helps in reducing the congestion that exists in the use of only one channel, as found in the conventional system (Kumbhar et al. 2016, p. 3). As a result, there is efficiency in communication. Some agencies have moved to the use of a trunked system, allowing them to increase the capacity of the number of users on a system and improve the interoperability potential.

The voice emergency services in the LMR systems work as follows. The individual responsible will usually dial the phone number for the emergency services, in this case, 911, on their mobile handset. The device will recognize the dialed number and disable any other services temporarily to prevent interference during the emergency call. The call will then be forwarded over to the cellular network with signaling that indicates to the personnel that the incoming call is one that requires an emergency response. The network then allocates the call for processing and, using the identity of the device's site, the appropriate emergency services network is selected. The process will also involve the originating networks, which are the dispatch centers that are in charge of obtaining the address of the caller with high accuracy. A communication pathway is usually established between the public safety answering point and the device that helps the receiver analyse the nature of the emergency. The personnel will then usually get the address from the caller if possible or determine the location through the use of the network.

5.2.2 Services Offered by LMR Systems

In public safety, one of the provisions of LMR systems is mission-critical voice communications. LMR systems help in providing the public safety responders with critical voice communication and also offer the best radio frequency coverage within the given geographic location of their responsibility. These systems have been designed in a way that allows the agencies to meet their mission-critical tasks by supporting time sensitive and lifesaving tasks (DHS 2011, 2016, p. 3). They include different elements such as high-quality audio, group calling capabilities, voice setups, and also guarantee priority access to the end users of the system. Furthermore, their infrastructure allows for prolonged operations in situations that occur in harsh and rigorous environments with easy accessibility and availability. Some of the system capabilities include dedicated channels, emergency alerting, encryption, priority access, and reliability in the networks (DHS 2011, 2016, p. 3). While the voice services can be offered through other types of system, none of them are able to guarantee the reliability and control that is demanded when it comes to communication in critical missions as well as public safety LMR systems (DHS 2011, 2016, p. 3). The LMR systems provide a wide range of capabilities that are of quality and are reliable and easily accessible to the users when it comes to priority communication during emergencies (DHS 2011, 2016, p. 3). Recently emphasis has been placed on the use of LTEs when it comes to communication in the public sector (DHS 2011, 2016, p. 3). However, the recommendation is that there is still the need to maintain and advance the LMR systems despite the integrated new technology present.

5.2.3 Adoption of Advanced Technologies to Supplement LMR

Public safety in the United States is a critical function that recognizes the need to have other means of communication other than LMR. It is often successful through the incorporation of many technologies that vary in terms of standards, capabilities, and requirements. LMR systems are supplemented with wireless broadband services and commercial cellular data to help in elevating them (DHS 2011, 2016, p. 2). None of the other technologies is meant to replace each other but they are meant to supplement the capabilities that are already in place in case the primary means of communication fails. They help act as backups and secondary communication systems and are sustained within the public safety agencies on a local and federal level to help ensure efficiency in mission-critical voice communication. Public safety agencies often combine largely commercial networks to help support the response efforts and help perform other services as well such as text messaging, transmission of low-resolution images and license plate queries (DHS 2011, 2016, p. 2). These data solutions supplement the LMR systems but are limited in their ability. Other options that are used to supplement LMRs are the wireless systems that enhance the responses through improving speed and capabilities at which the critical data is being received.

To help ensure interoperability, P25 standards are used in the LMR systems. The systems offer the manufactures the required specifications when creating LMR systems. The main aim of these standards is to ensure interoperability and compatibility with other systems. While it is not a requirement for the different agencies to invest in P25 only equipment, many of the federal grant agencies do not approve of other equipment (DHS 2011, 2016, p. 3). Before P25 was introduced, the manufacturers in charge of building the systems would normally do so based on their own functionalities and features (DHS 2011, 2016, p. 3). These systems would then be owned and operated by public safety agencies. However, there were challenges in these systems in their operability (DHS 2011, 2016, p. 3). The different systems from the vendors would sometimes fail to be compatible, greatly inhibiting interoperability (DHS 2011, 2016, p. 3). In order to prevent this, the government encouraged the agencies to purchase equipment that complied with P25 standards (DHS 2011, 2016, p. 3). The P25 standards helped ensure compatibility in operations, allowing effective intercommunication with various agencies and jurisdictions in the event that emergency respondents needed to communicate during times of disaster, day to day incidences and large-scale emergencies (DHS 2011, 2016, p. 3).

5.2.4 Trunked Digital Network

Within the EU, the two technologies used in public safety are TETRAPOL and terrestrial trunked radio (TETRA).

5.2.4.1 TETRAPOL Communication System

The TETRAPOL system is known around the world and is responsible for facilitating communication through digital radio options for public security and military operations in many countries within the EU (Bârcă 2017, p. 185). The system uses frequency division multiple access technologies (FDMA) with a narrow bandwidth in the frequency range of between 70 and 520 MHz (Bârcă 2017, p. 185). The system complies with the professional model radio to ensure that it provides the efficient radio coverage required in both densely populated areas and those that are inaccessible (Bârcă 2017, p. 185). It also offers an infrastructure that can support interconnection across the borders.

5.2.4.2 The TETRA Communication System

The TETRA system is also known as terrestrial trunked radio and is a widely used digital radio system. The standards that it operates on have been approved by the European Telecommunication Standard Institute as an official radio communication channel across Europe. It offers both professional and regular cell services and its unique combination has enabled its use in public safety (Bârcă 2017, p. 186). It contains a unique combination of mobile phone, data services, and group communication services that can be especially convened for use by the authority only. In TETRA systems, the radio channels are centralized, and the systems normally automatically assign the different channels to the users at the beginning of any call (Bârcă 2017, p. 186). It is considered more efficient than analog because of its channel width, which can take four communication channels (Bârcă 2017, p. 186). The first version of TETRA was known as the open systems and was developed by ESTI in order to facilitate functionality and rapid adoption by the different national administrations (Bârcă 2017, p. 186). Efforts on standardization are still ongoing due to new advancements in technology allowing interconnectivity and interoperability among the features. The first version included voice and data but became inefficient with the increase in demand from the users within the public safety agencies due to the technology evolution (Bârcă 2017, p. 186). The aim of interoperability in TETRA systems is to be able to allow the flow of critical information from any sector. It helps in improving accuracy and timeliness by reducing the number of people who may have committed errors and the duration of the circuit so that there is as little interruption as possible during communication. In Europe, standardization is a key concept for those in charge of public safety (Bârcă 2017, p. 186). Interoperability between the different networks in public safety creates shared networks that allow for flexibility and reliability in the use of the system for emergency responders. In the technical development of the system, there is the aspect of standardization, interoperability, manufacturing, and final testing (Bârcă 2017, p. 186). It allows countries within Europe to use them in public safety on a wide range of communication and operational opportunities (Bârcă 2017, p. 186). It helps in providing secure data and voice transmission that allows communication in all the public safety agencies within a single system.

5.3 Is 4G LTE Forming a Good Enough Basis for Public Safety Implementations?

5.3.1 Multi-Path Approach and the Convergence of Mission-Critical Communication

4G LTE does form a good enough basis for public safety implementation. This technology has come with improvements and new content within it that have been able to ensure that it is a more efficient means of public safety communication compared to the current systems in place (TCCA 2016, p. 9). Some of the essential benefits that come with it include a significantly improved bandwidth that allow it to transmit data faster, with shorter roundtrips delays that allow systems to be more responsive, improved quality of services are offered and the network architecture is simplified and therefore will lead to fewer costs associated with implementation for the mobile network operators (TCCA 2016, p. 9). In times of emergency, access to a reliable communication is crucial. Unfortunately, it is currently hindered by the divergent signaling standards in place that have failed to facilitate high-speed data performance that is necessary to support the applications that are relied on by public safety agencies today. 4G LTE offers these services and helps in catering to these demands (TCCA 2016, p. 9). For public safety agencies, there is a need for an integrated and unified broadband infrastructure that features the strengths of the current and new technologies with continued access as advances made to the commercial cellular network continues.

4G LTE is a broadband technology that is able to facilitate high data rate applications that are not possible in the current LMR systems. It has been developed by 3GPP as a form of 4G mobile network communication with the main goal of increasing capacity and facilitating high-speed rate of data transfer in wireless networks (Lynch 2016, p. 15). In LTE, one of the biggest benefits to public safety has to do with its data connectivity. The previous public safety networks usually supported low data rates and usage with many of their aspects only allowing the enabling of voice-oriented features. 4G LTE systems have the ability and capacity to transfer high volumes of data immediately (Lynch 2016, p. 15). This means that there can be the ability to send high-quality images thus enabling the use of other services such as live streaming in times of emergencies or the use of the mapping software to be able to direct the mission-critical resources. A database query in the LTE also delivers rich and easy to read reports that contain images such as video clips and mug shots. They are more efficient than the small paragraphs of text-only information that is presently being provided by LMR systems (Motorola 2012, p. 6). High-speed broadband communication helps make a difference and can be applicable in many cases. For police officers, high-speed data helps provide access to video of an ongoing crime and allows them to be updated as they are on their way to the scene of the crime (Motorola 2012, p. 6). It allows them access to life-saving information on how to best approach the situation with minimal damage. The program will also remove the traditional amber alerts issued in cases of missing children and instead of just providing the alert, helps through giving a detailed description of the car and access to high-resolution images of the child and the perpetrator (Motorola 2012, p. 5). This information is accessible within minutes, to not only the first respondents but also to municipal employees. It acts as an alternative to calling in cases of accidents as the video surveillance camera are able to immediately catch the accidents and alert the command center immediately. This, in turn, allows improved workforce productivity and improved communal responses to threats in society (Motorola 2012, p. 6). LTE will allow the improvement of emergency responses through different applications that focus on dynamic mapping, traffic flow, and weather.

5.3.2 Technical Aspects of LTE

The technical work to help in the production of 4G LTE standards to be able to support public safety communication was done in accordance with 3GPP. The main objectives of the 3GPP include helping preserve the strength of the LTE while adding other features in public safety and to maximize other technical abilities that provide the most cost-effective solutions in public safety communication (Sharp 2013). In 4G LTE, there are a few main areas that have been agreed up to help facilitate public safety applications according to 3GPP standards. One centers on the proximity services and enabling the optimization of communication in regard to physical proximity while the other looks at the group call systems in place that help support the fundamental requirements, which allow for a more dynamic and efficient group communications operations associated with these two main areas (Sharp 2013). Security features are also needed to help in protecting the systems from fraudulent use and eavesdropping. The proximity services to be offered consist of two main elements. One is the facilitation of direct communication between the users and the network assisted discovery of the individuals in public safety who have a desire to communicate when in close proximity to each other with or without supervision (Sharp 2013). Direct communication deals with how a radio network can be established between the users without transition through a network. This helps in saving network resources and also allows the public safety communication of the users outside the network coverage (Sharp 2013). In the commercial area, proximity can help support the transfer and sharing of information between the devices (Sharp 2013). The definition of proximity services involves features that can be applied in the public safety spectrum (Sharp 2013). In these features, the mobile acts as a relay to one another and provides access to many of the network services when outside the network range. The other features allow communication without the network facilitating direct communication.

Group communication within the standards is one of the most critical requirement and allow the cooperation and coordination of the first and second respondents in public safety. It includes broadcasting of data to all the relevant personnel providing information on the current situation (Sharp 2013). It also allows communication between closed groups through LTE features such as multicast services to help in supporting all the requirements.

5.4 Is It Better to Wait for 5G Before Starting Public Safety Implementations?

While 5G is a better option in public safety implementations, 4G also serves the required functions in public safety and there exists the specified spectrum that can support its function. 5G for public safety currently has several constraints that may take some time to sort out. As a result, the use of the next best available option will serve to improve current public safety communications and help in creating advances that can be supported when the 5G network begins to be implemented. One of the constraints related to 5G surrounds its spectrum (Wang et al. 2017, p. 28). This is because the current public safety applications used by FirstNet in the United States have their own assigned spectrum. The technology may be useful and desirable, but it is not economically feasible to employ until it is perfected while ignoring the current applications that can improve the services at the present (Wang et al. 2017, p. 14). Many states and municipalities are also limited in resources, therefore they lack the ability to be able to sustain 5G as a supplement within the existing system working in the direction to completely eliminate them. Looking at the positives, 5G is a better option for the public safety communication implementation because it allows for higher speeds, is more reliable, and has the capacity (Wang et al. 2017, p. 18). The additions from 5G technology can be applied to improve public safety in different ways. The speed can be used to help in vehicle-to-everything communication and facilitate safety on the roads. It also allows vehicles to have self-reporting mechanics in regards to traffic incidents and other accidents (Wang et al. 2017, p. 30). The very fast vehicle-to-vehicle communication that 5G facilitates allows emergency respondents such as police, firefighters, and ambulances to become notified in cases of emergency. The capacity that 5G provides also makes it a better public safety implementation than the 4G networks. They are used in a variety of situations, including in the support of high-quality video and sharing of large files that contain important data that help provide emergency services (Wang et al. 2017, p. 30). It is most likely that 5G will act as a supplement within public safety communication systems that are currently in place and are being built. Another positive that makes it better in public safety implementation is in regard to high priority situations. It has the next generation network priority service (NGN-PS) that influences decisions during high priority situations (Wang et al. 2017, p. 28). The NGN-PS helps obtain resources to resolve critical events (Wang et al. 2017, p. 38). The platform is currently interfaced with the 4G network, which has made it more flexible and able to provide a wide range of capabilities that can be used to expand priority services (Wang et al. 2017, p. 28).

5.5 Will 5G Offer a Better Service than 4G for Public Safety?

Any new type of 3GPP release is always able to offer better services, improvements, and capabilities when compared to the previous one. In this case, 5G offers better services compared to 4G in public safety due to the elements within that surround wireless innovations. 5G wireless networks are helping to address the evolution of the Internet of Things (IoT). The main difference compared to 4G is that it helps in data speed improvements and goes beyond to address the new IoT and critical communication used in the latest technologies (TCCA 2016, p. 6). The network helps provide real-time interactivity in services such as the cloud, a key factor in the success of self-driving cars, which is a new trend that is being analyzed as part of public safety. Unlike the current IoT that makes performance trade-offs in order to obtain the best from the wireless technologies currently in place, 5G networks will be able to allow a fully interconnected world, bringing the level of performance to the expected standards and ensure efficiency in public safety communication (TCCA 2016, p. 5).

Rather than just 4G with added features, the 5G wireless network will also create a new paradigm between people and machines of interconnectedness. This is referred to as "the Internet of Things" and will affect the whole infrastructure including the start grid and critical infrastructure with public safety communication (TCCA 2016, p. 11). It will also support a much more advanced version of the current multimedia applications such as virtual reality. This will, in turn, lead to an increase in data usage that is far beyond that of the current network capacity. The robust infrastructure will be reliable and lead to significant improvements in public safety if implemented.

5.5.1 The Internet of Things and 5G

One of the main reasons why 5G will be offering better services than 4G is the IoT. The IoT refers to the processes, people, and data that come together to make connections in the network that are more relevant and valuable (TCCA 2016, p. 11). The evolution of the internet has allowed changes in technology that require a faster and more reliable means of communication from that of 4G (TCCA 2016, p. 11). 5G can facilitate this through faster data transmission, which helps to simplify the shared device management. The

IoT component is comprised of a network of physical objects that are accessed through the internet and can communicate data (Gopal et al. 2015, p. 70). The technology helps through allowing users to take action based on the intelligence data given. The development of the IoT has helped expand the scope and capabilities of the cameras connected that have become part of the intelligence platforms that help deliver data for operation efficiency, forensic investigation, and situational awareness (Gopal et al. 2015, p. 69). The cameras are able to provide superior insight into all the application areas from security, safety, and the operations within the public safety communication. With the use of 5G in IoT platforms, the emergency respondents can get sensitive information over a secure network environment through their mobile phones. This is because the IoT intelligence and the presence of next generation networking can help connect public safety to their data through the use of a single secure solution (Gopal et al. 2015, p. 70).

The new phenomenon of the IoT also helps provide an advantage of using 5G technology through its ability to take simple automated tasks and enabling them to communicate their data via the connected devices. Eventually, the tool will assist mobile devices to be able to interpret data on location and preferences when it comes to public safety without requiring any input (5gAmericans 2017, p. 20). 5G will have the ability to help in the fast and resilient access through the internet regardless of the geographic location (5gAmericans 2017, p. 19). 4G works only with the availability of the raw bandwidth creating a difference between the two (5gAmericans 2017, p. 20). 5G technology is also being currently designed from the ground up to help in supporting some of the machine type communication compared to 4G LTE, which requires the incorporation of the variant called the machine type communication to help in IoT traffic (5gAmericans 2017, p. 20). 5G is also better than 4G because it does not operate as a monolithic network and is built around the integration of different technologies from the previous generation starting from 2G to M2M (5gAmericans 2017, p. 20). It will thus be able to support a wide variety of applications in public safety communications including that of the IoT, and wearable devices and sensors when it comes to the emergency responders.

5.5.2 5G Technical Aspects

As the digital industry moves toward a more heterogonous technology that contains a multilayer network, which consists of microcells, low powered small cells, and relays supported through digital connectivity, 5G is the most technically feasible technology that is able to meet these needs in public safety (Wang et al. 2017, p. 28). The next generation technology consists of a collection of technologies, and utilizing the already existing fixed networks in place for wireless connectivity will help cater to the different elements, including that of the spectrum, high performance, and stable infrastructure to offer emergency responder services (Wang et al. 2017, p. 28). Network densification is the main element that has driven public safety toward the building and implementation of 5G technology within the public safety system. This is especially the case with the problem of the high-frequency bands that are widespread and that have poorer characteristics. The 5G network has its own high-frequency spectrum that can support the 700 MHz required in public safety through its ability to reach above 6 GHz and its use of the millimeter wave spectrum (Wang et al. 2017, p. 35). The millimeter wave spectrum has the potential to offer multi-GPS at a much lower marginal cost than the existing and previous technologies used within public safety due to its ability to be able to allocate more bandwidths with its frequencies (Wang et al. 2017, p. 28). They are more applicable to local hotspots rather than wide area coverage.

5.5.3 5G Network Costs

Cost modeling is a method that allows comparison of the different data traffic demands and the network costs that are associated with the technologies (Oughton and Frias 2017, p. 2). In 5G, the costs depend on different factors in the public safety system including the density, base station prices, and the periodic interest rates (Sharp 2013). A comparison of the costs helps in determining the efficiency of using 5G as a better option compared to 4G LTE technology. 5G integration of the millimeter wave base spectrum can be used in the deployment of emergency services to high demand areas to make it a more effective system for ultra-dense small cell deployment (Marchetti 2015, p. 8). The largest costs associated with the use of the 5G are related to the individual unit prices of the base station that are required to help in delivering the 5G network of a required capacity coverage. This, however, can be solved with time through the application of the software defined networks that are currently still at the nascent stages of development (Oughton and Frias 2017, p. 2). The evolution of these networks within radio access and the evolved core will allow a significant reduction in the building up of the infrastructure of 5G networks compared to those of traditional systems (Marchetti 2015, p. 10). Network infrastructure, in general, has very high fixed costs when it comes to implementation, placing a great burden on the scale of economies and the population density in public safety communication. This infrastructure affects decision making when it comes to the application of 5G over 4G networks due to the risks associated and the extensive resources required that are not easily available because of the public safety budget that is in place (Marchetti 2015, p. 9). One way that has been suggested to over this is through the use of the open deployment cells serving the subscribers in public safety to be able to create a shared infrastructure approach that would allow the encouragement of market entry, thus bringing the networks closer to the masses (Oughton and Frias 2017, p. 2).

5.5.4 Key Corner Cases for 5G

The key corner cases for the use of the 5G over 4G and other previous technologies within public safety systems lies in its foundations. One of these is enhanced mobile broadband, primarily referring to video downloads (Bishop 2016). With the recent increase in video traffic and the need to add more of the radio frequency spectrum, 5G networks come in handy for emergency respondents to help support this function. The second corner case is related to massive machine type communication, which is used to help in the connection and communication with many other devices (Bishop 2016).

Low latency communication (LLTC) is another corner case for the 5G network over the 4G because it has the ability to facilitate fast communication within the public safety networks that can be integrated with wireless communications to allow the insertion of emergency messages (Bishop 2016). The ultra-reliable LLTC is very crucial in vehicle-to-vehicle communication to help serve the different emergency needs such as collision avoidance. The current 4G LTE has only 10 ms of one-way latency (Bishop 2016). The goal of the 5G networks is to be able to reduce this number to only 1 ms to help in faster communication. Research has found that a good 4G LTE network usually has about 50 ms of latency that reduces the speed of communication within the public safety sector (Bishop 2016). The 5G network has been moved toward a more network edge to be able to increase functionality and reduce the end to end latency to 5 ms. In the application of public safety, latency can be explained in vehicles and how they are able to communicate emergency information to its users (Bishop 2016). Use of 5G can allow the communication to the user within five seconds to be able to make them aware of when it is time to stop.

5.5.5 Localization in 5G Networks

5G networks will also support network-based localization, an advantage that was lacking in the previous technologies used in public safety. The process allows these dimensional spaces with an accuracy of less than 1 m or in 80% of the occasions (Zhang et al. 2017, p. 28). It will allow the next generation to be able to benefit from localization in the process of wireless network optimization. In the previous technologies, the multi-tokens (MTs) were not able to support accuracy when it came to localization (Zhang et al. 2017, p. 28). Network-based localization is thus effective to helping to provide a reliable and faster emergency response. The structure is being incorporated in the early stages of the 5G development process to be able to allow fast location area communication, so the economic services can be quickly received (Zhang et al. 2017, p. 28). Some of the innovative technologies that are being applied in the localization include that of smaller cells, higher frequency bandwidth, higher throughput multi-token MAC (MTM), and device-to-device communication. Smaller cells will be the nested cells in 5G such as picocells under the range of 100 m and Wi-Fi in a range that helps to foster the short time required in emergency response (Zhang et al. 2017, p. 28). The MTM and device-to-device communication distinguish 5G from the previous networks in public safety because it helps in facilitating high accuracy within the localized areas. Through the channel estimations of the MTs, the internode found can be measured and be extracted with the location information (Zhang et al. 2017, p. 28). Additionally, the device to device combination allows the free exchange of the data, which can be applied when distributing the shared channel information.

5.6 What is the Linkage Between 4G–5G Evolution and the Spectrum for Public Safety?

5.6.1 The Linkage Between 4G-5G Evolutions

The linkage between 4G and the 5G is as a result of 3GPP releases and standards that have helped in the evolution of technology through the different generations to reach 5G. The 3GPP has a long history of coming up with and defining the different technologies that are used within public safety communication. This evolution includes from the global system for mobile communication (GSM) to high-speed packet access (HSPA), 4G LTE and finally to the upcoming 5G technology (Borkar et al. 2012, p. 10). The process of adding new features to the previous releases and using their own applications as the basis for new technology has provided a connection between all the releases that are able to operate based on the spectrum frequency that is a commonality in public safety systems. The success of the 4G LTE in public safety can be traced to the definition of Releases 8 and 9 by the 3GPP that has led to LTE advancements in Releases 10 through to 12 (5gAmericans 2017, p. 214). Release 13 of the 4G LTE in mission-critical communication completed in 2016 has key enhanced features added to it to help in providing support for active antennas, aggregation of LTE and WLAN, LTE in an unlicensed spectrum (LAA, license assisted access), and low power to wide area coverage for IoT applications through the use of narrowband IoT (5gAmericans 2017, p. 214). They also enhanced the previously introduced LTE technology from Release 12 such as carrier aggregation (CA) and dual connectivity, advanced multiple-input multiple-output (MIMO), self-optimizing network (SON), proximity services, and device-to-device communication for public safety. As technology evolved, further advancements were made in Release 14 to the MIMO, CA, and LAA (5gAmericans 2017, p. 214). With the rollout of 4G LTE and the evolution of the wireless industry, the idea for the development of the next generation technology came about. This led to research and the building of the infrastructure of 5G technology in order to support the wide variety of frequency ranges within the public safety systems (Borkar et al. 2012, p. 10). The approach taken to initiate the 5G technology began in two phases. The first phase involved focusing on the 5G technology with respect to the next radio and generation systems architecture by the year 2018 (5gAmericans 2017, p. 156). The second phase included meeting all the needs set out in the 3GPP standards by the year 2019. This shows the influential force that the 3GPP has in creating a link between the different generation technologies and driving the evolution of LTE technology to support the introduction of the 5G network in Releases 14 and 15 (5gAmericans 2017, p. 156).

Ensuring interoperability with the previous technology used in communication is one of the key principles in the 3GPP standards (Borkar et al. 2012, p. 11). The main aim of the 5G was to ensure connectivity for any device or application that may be applicable to benefit the public safety agencies. The portfolio of connectivity and access solutions will help in addressing public safety demands while meeting the requirements of mobile communications beyond the year 2020 (Borkar et al. 2012, p. 11). The specifications of the network include features such as the new and flexible air interface, NX, which help in the targeting of the scenarios that involve the mission-critical and real-time communications based on the requirements that cater to latency and reliability. The use of the 3GPP is intended to help in the support of the massive machine connectivity that is a presence in wide area applications that help provide the deep coverage and high capacity for a large number of the connected devices (Borkar et al. 2012, p. 10).

5.6.2 Spectrum for Public Safety

In public safety, the spectrum is considered one of the main issues in the 5G vision. Different international bodies have found that the spectrum for public safety operations can be found in the 700 MHz band. However, different countries determine how much and how they will apply to this band. The 700 MHz will currently be implemented through the 4G network until the completion of the next generation technology that is being updated and standardized by the 3GPP in Release 15 (LaMore 2008, p. 4). One of the goals in the use of the 700 MHz band was to be able to help increase the capabilities and the amount of spectrum in public safety (LaMore 2008, p. 5). After the different disasters that occurred within developed countries such as Hurricane Katrina, many government bodies including that of the United States decided on the need for a clear public safety network that is capable of supporting emergency communications. This was found in the 700 MHz band spectrum, which has been applied and restructured over time to help in supporting the different bandwidth applications in order to improve interoperability (LaMore 2008, p. 5). In the United States, the shift to the use of the 700 MHz band initially began when the switch occurred from analog to digital in terrestrial communication in the year 2009. This 700 MHz bandwidth developed as a result of the switch to the lower bandwidth found in digital technology that was able to free up spectrum that can be allocated to other uses.

The 700 MHz contains two separate bands; that is the lower and upper 700 MHz band (LaMore 2008, p. 4).

The lower band is separated into five different blocks to facilitate the transmission of data (LaMore 2008, p. 4). The first three A, B, and C are usually brought to work together in pairs to help in their support of two-way communication (LaMore 2008, p. 4).

In the set-up of the band, there is the upper half that mainly supports the transmission of signals from the towers to portable devices (LaMore 2008, p. 5). The lower is responsible for working in reverse in that it transmits from the portable devices back to the tower networks (LaMore 2008, p. 5). The blocks C and D remain unpaired and are used within the spectrum to help in one-way transmission (LaMore 2008, p. 5).

The upper 700 MHz band is also broken down into five different blocks whereby blocks A to D are re-utilized for commercial services while block E is designated to facilitate communication that supports public safety. Blocks A and B were initially used as guard bands within public safety, which meant they had to adhere to the strict standards in how to reduce interference with the neighboring frequencies (LaMore 2008, p. 6). They were not allowed to operate in the cell-based infrastructure, which limited their capability with the advancements in technology that centered around mobile-based developments However, after the restructuring, only block B remained subject to these requirements while A shifted and adopted features that supported a cell-based infrastructure. Block D has only one license and is registered nationwide for the entire block. It is also the only feedback that is solely designated to serve public safety functions (LaMore 2008, p. 4). All these blocks are structured in a way that they are able to offer support to public safety devices.

In the use of the 700 MHz band to increase the amount of spectrum for public safety use, 24 MHz is solely designated to cater to the national, local, and regional agencies responsible for public safety applications (LaMore 2008, p. 2). The 24 MHz is broken down where 5 MHz is allocated for national networks while the rest is allocated to the regional and local agencies to help support voice communication and low-speed data (LaMore 2008, p. 6). The nationally controlled broadband network fills the large gap that is found in the public safety network, allowing it to be used in any region around the country. The need has been shown to be clear in many instances such as the 9/11 terrorist attacks in the United States (LaMore 2008, p. 2). The creation of one major broadband use within the spectrum helps in the facilitation of emergency services, coordinate of security, and deployment of secure operations easily through the shared network around the country.

In the 700 MHz band plan, there are different channels set aside to help in the low power usage during emergency response. The first channels are paired to help in facilitating low power use in on-scene incident response through the use of mobile and portable devices that are subject to the approval of the public safety agencies governing bodies (LaMore 2008, p. 2). The transmission power in these channels is limited to a certain number of watts (LaMore 2008, p. 2). The second channels are paired to help in the operations in either analog or digital networks. To facilitate analog implementation, the plan allows the use of up to 12 kHz bandwidth to help in creating the on-scene temporary bases and mobile relay stations that can facilitate communication and emergency responses (LaMore 2008, p. 2). In its digital implementation, there is the allocation of different bandwidth frequency pairs that have the potential to support multiple low power applications in the same frequencies. The plan also ensures that there are minimal inferences between the frequencies that may result because of the power restrictions.

One of the important elements found in the use of the 700 MHz spectrum is that of the narrowband segment that consists of various channels operating in pairs (Federal Communications Commission [FCC] 2017). The segment is subject to the usage efficiency requirements set within the spectrum and are divided to help in facilitating different services within public safety networks. One of the services it offers is the narrowband interoperability channels that have been designed to help in the interoperability use of data and information between different channels (FCC 2017). The second is the general use channels that have been designated for general use for public safety. Next is the general use (former reserve channels) designated to support the temporary, mobile infrastructure in the incident area to help in assisting emergency teams in the efforts of response and recovery (FCC 2017). The state channels help provide support to each state or territory while that of air–ground channels are reserved for use in air-to-ground communications by ground base stations or low altitude aircraft. Thus, the allocation of 700 MHz in public safety through narrowband channels has been able to ensure that public safety agencies and applications are able to take advantage of the more efficient technologies, leading to a reduction in channel width and the ability to assign additional channels if the need arises (FCC 2017). Narrowband helps in increasing the spectrum and its efficiency including allowing room for interoperability with any other future added features.

5.7 Conclusion

Ultimately there have been significant changes to public safety systems in the world today, a trend that is unlikely to stop anytime soon with the introduction of 4G LTE and the development of 5G networks. The existing public security network is based on systems that have been in use for decades, facilitating mission-critical communication within the public sector. These systems differ from country to country but work with the same goal of ensuring efficiency in public safety. The development of 4G LTE has been able to transform public safety networks in many countries due to its added features that allow video data and faster responses to emergencies. They are able to offer services that were lacking in the previous technologies, allowing public safety to access data connectivity. With the development of the 5G network, many have wondered whether public safety should wait until it ready for use instead of using 4G LTE. However, 4G LTE is compatible with the current frequencies in the market ensuring its availability within different states. The spectrum has yet to be fully developed and available making it economically unfeasible to implement. 4G can be provide services that improve emergency responses. However, 5G is a better option compared to that of 4G due to its added features that allow it to be faster than 4G due to data improvements, offer better services, and have more capacity to support users.

References

5GAmericas (2017). *Wireless Technology Evolution towards 5G. 3GPP Release 13 and Release 15 to Beyond*. Bellevue, WA: 5GAmericas.

Bârcă, C.D. (2017). TETRA system – open platform – interoperability and applications. *Journal of Information Systems & Operations Management* 11 (1).

Bishop, Don. Why Public Safety Communication needs 5G. *AGL Magazine*. (2016).

Borkar, S. et al. (2012). *Application of the Upper 700 MHz A-Block to Public Safety*. Roberson and Associates, LLC.

DHS (2016). *Land Mobile Radio (LMR) For Decision Makers*. Washington, DC: Department of Homeland Security (DHS) Office of Emergency Communications.

DHS (2011). *Public Safety Communications Evolution*. Washington, DC: Department of Homeland Security (DHS).

Federal Communications Commission (2017). *700 MHz Public Safety Spectrum*. Washington, DC: Federal Communications Commission.

Gopal, B.J. (2015). A comparative study on 4G and 5G technology for wireless applications. *IOSR Journal of Electronics and Communication Engineering* 6, 1: 67–72.

Kumbhar, A., Koohifar, F., and Guvenc, I. (2016). *A Survey on Legacy and Emerging Technologies for Public Safety Communications* (ed. B. Mueller). Motorola Solutions, Inc.

LaMore, Adam. 2008 The 700 MHz Band: Recent Developments and Future Plans https://www.cse.wustl.edu/~jain/cse574-08/ftp/700mhz.pdf

Lynch, T. (2016). *White paper: LTE in Public Safety*. IHS *TECHNOLOGY* https://technology.ihs.com/580532/whitepaper-lte-in-public-safety.

Marchetti, N. (2015). Towards the 5th Generation of Wireless Communication Systems. *ZTE Communications* 13 (1): 11–19.

Motorola (2012). *The Future is Now: Public Safety LTE Communications*. Chicago, IL: Motorola.

Sharp, I. (2013). *Delivering Public Safety Communications with LTE*. Sophia Antipolis: 3GPP.

Oughton, E.J. and Frias, Z. (2017). The cost, coverage and rollout implications of 5G infrastructure in Britain. *Telecommunications Policy* 42: 1–17.

Wang, T., Li, G., and Huang, B. (2017). Spectrum analysis and regulations for 5G. In: *5G Mobile Communications*, 27–50. Switzerland: Springer International Publishing.

TCCA (2016). *4G and 5G for Public Safety: Technology Options*. Newcastle upon Tyne: TCCA.

Zhang, P., Lu, J., Wang, Y. et al. (2017). Cooperative localization in 5G networks: a survey. *ICT Express* 3 (1): 27–32.

6

Roadmap Toward a Network Infrastructure for Public Safety and Security

6.1 Introduction

There are a number of emergency services that utilize communication pathways between personnel responding to emergency events and/or preparing to respond to events. The various communication pathways used by emergency response personnel share critical information related to safety and security of both personnel responding and those personnel who are receiving the emergency aid. The importance of network connectivity and the ability to effectively and reliably communicate between responding agencies and individual responders is understood, without the need for much clarification.

6.2 Evolution Toward Broadband

In terms of quality critical communications, public safety organizations have reserved the best networks whose efficiency is based on voice services. Services in public service organizations are conveyed through digital technologies with dedicated network implementation that ensures that they remain effective. Government-controlled operators ensure that the dedicated approach ensures that there is certainty of operations in providing critical services to citizens. Over voice and short messaging services, high availability and reliability have been achieved, enhancing the operations of the public safety operations. Technology has however evolved, and if private organizations have the ability to deploy services that ensure they can send video on demand, public safety organizations should also have dedicated networks that allow for the deployment of the same. Current professional mobile radio (PMR) standards do not have the capacity to guarantee that, using the existing systems, the public networks will have the capacity to support, in addition, critical applications that are synonymous with today's mobile computing technology. Obtaining the reconfiguration of the available networks to support broadband would rely on:

- Improved operational efficiency. Mission-critical field services, as well as public safety operations, should leapfrog the available efficient private networks through the application of new applications that supports communication amongst the providers. Efficiency in the model would be achieved by ensuring that there is support for remote data access, instant image transfer, video on demand as well as mobile office interactions that enhance connectivity and full productivity, even if responders are far from their areas of operations.

- Increased demand for public safety. The agenda of government concerning safety has been enhanced in the recent past by the issue of global terrorism that has emerged as a public safety concern. Maintaining safety is profoundly important not only to the first responders but also to the law enforcement officers and citizens at large. Some core efforts like maintaining safety across borders require intensive technologies that would for example beam the images of movement across borders on demand. This is important in determining the right approach to take in apprehending bad elements with bad intentions before they have the capacity to breach the peace.
- Cost reduction. Governments have the added civil responsibility of ensuring that resources are adequately used in the best way possible without wastage. Maintaining older technologies is an expensive undertaking compared to employing technologies embraced by the private sector. Improving the efficiency and speed of data transmission by employing the use of broadband services significantly reduces the running costs associated with maintaining older technologies. Monthly operating expenses (OPEX) operational fees are for example equivalent to what technologies like capital expenditures (CAPEX) would use for the duration of one month.

6.2.1 Existing Situation

Existing public safety telecommunication networks also regarded as public protection and disaster relief (PPDR) are largely based on standards established in the 1990s. Europeans nations use the TETRAPOL standards while others use TETRAPOL. The standards in use in North American nations is P25, which offers enormously reliable narrowband used for supporting voice communication from person to another or in group calls, and with limited capacity for data and messaging (Munir et al. 2015, p. e3008). Operations of the allocated frequencies are below the ranges of the 400 MHz frequency spectrum that are ideal for low bit rate services.

The number of users that access the networks for communication is limited, with police units, border patrol, ambulance, fire brigade, and national security networks using them at different frequencies. Across nations, the frequencies vary as the networks have additional safeguards in place intended to limit the number of users to particular service organizations. Unlike the commercial networks that are scattered and varied based on location challenges, the public safety telecommunication services have a wider coverage across different locations. In networks like TETRA, for example, there is bidirectional authentication required between the personnel using the services and the base stations (Guo and Rigelsford 2013, p. 2852). The existing systems are also different compared with the commercial networks in the symmetric data transmission patterns with down-links almost matching the uplinks. In commercial networks, the traffic is asymmetrical with downlinks consuming substantially more traffic compared to uplinks. The commercial networks have the advantage of having the capacity to upload or download almost any type of file without constraints. Evolving the systems to have the capacity to send large data files will be the core consideration with any effort toward enhancing the capability of existing public safety networks. Interoperability with the commercial systems would be a critical addition to the new envisaged network to have the ability to connect to the Global System for Mobile Communications (GSM)/Universal Mobile

Telecommunications System (UMTS)/Long Term Evolution (LTE). A TETRA device connecting with the commercial networks would, for example, allow for the upload of videos, files, and images to the base systems.

6.3 Requirements for Public Safety Networks

The overall shift in technology to embrace digital telecommunication networks and use them as enablers for increased interaction and exchange of information make it imperative that public safety networks have the same features to be relevant to societal demands. It is no longer sufficient for public safety networks to have infrastructure that only supports voice and short message services (SMS) while private networks have the capability of uploading and downloading video on site. The emergence of social media networks and the wide usage of the 3G, 4G, 5G, and LTE networks make it easy to upload voice and video, which would help the emergency services develop situational awareness quickly and reliably. Having reliable network connectivity for emergency systems with the capacity of streaming high-quality video and sound content would, for example, play a vital role in saving responders lives by replacing them with smart robotic systems that can access volatile environments. Such devices would need wireless broadband to improve the visual and device to device connectivity (Usman et al. 2015, p. 1652). Amongst other advantages, integration of PPDR to LTE or 5G systems would enhance:

a) High-resolution imagery
b) Robotics and drone control
c) Video live feeds with capacity for quality streaming, uploads, and downloads
d) Full database access for internet, intranet, and organizational directories.

Compared to private data networks that support the public mobile connectivity systems, TETRA, TETRAPOL or P25 lack the capacity to guarantee sufficient cell throughput necessary in the worst conditions and upload the required information. Conventionally, such throughput is only in the range of 1–2 Mbps, which is impossible in the public safety networks but available in LTE based technology. The transmission links supported by the public safety networks that use time-division-multiplexing (TDM) links would have to be upgraded to support internet protocol (IP) based high capacity links. Changing from one established system to another requires the capacity to blend the old while testing the new, and broadband services should support the PPDR network narrowband and wideband systems for seamless integration. Such hybrid systems are, however, costly and redundant, but for a period of time are necessary to support the migration process. However, broadband based PPDR networks are necessary as a matter of urgency since the mobile technology networks are ever evolving and the blending phase may be left out in transition to a newer technology. Mobile based broadband network systems are dynamic and ever-evolving, requiring an element of a quick resolution to keep pace. The existing narrowband networks that today seem like a relic of the past only became a standard designed in the 1990s. Public safety networks are critical service systems whose availability and reliability must be guaranteed at all times requiring a different level of standardization to insulate them from any kind of eavesdropping, spoofing, or attacks intended to trigger a denial of service. Benchmarks and parameters necessary to guarantee such levels of system firewalling should be put in place before dual testing.

6.3.1 Network Requirements

Current drives to bring about the benefits of commercial broadband networks to mission-critical service providers have developed solutions "over-the-top" of existing public networks. Doing so is, however, not sustainable in the long-term since the public safety networks desire a level of sophistication that ensures they get priority in the systems (Tata and Kadoch 2014, p. 3). The networks should be managed in the same way that TETRA, TETRAPOL, and P25 works, giving priority bandwidths to public safety networks while ensuring that the communication is encrypted end to end. Even in extreme weather conditions, the dedicated broadband services for the mission-critical services must ensure:

a) Survival over multiple failures
b) Priority for mission-critical data
c) Offer the desired coverage and capacity
d) Maintain data integrity and ensure end-to-end encryption
e) Be interoperable with other networks to offer solutions when needed
f) Provide the desired support for devices and applications.

6.3.2 Priority Control

At times of emergency, the networks must ensure there is a priority over the traffic for commercial users both in admission control and traffic priority. The extent of prioritization is, however, both static and dynamic and may affect the manner in which services are offered. Static prioritization, for example, would be accomplished through offering voice calls emanating from dispatcher's priority over other calls (Fereidountabar 2013, p. 149). In the same vein, the different network requests would be offered with a level of priority with voice calls, for example, getting precedence over web browsing requests. Dynamically, however, the need to prioritize some requests over others would require the input of the concerned parties to understand the right algorithm to use and determine what to give priority. Priority control would be able to determine, for example, which emergency responders to accord priority in times of disaster based on location and other relevant factors even as all other responders access the network. The complexity of making decisions concerning which facet of the mission-critical service providers would require close collaboration between public safety stakeholders and the persons driving the innovation processes. Prioritization equally would need decisions on according public safety networks priorities, especially in areas where coverage differs according to different parameters. In the urban environment network traffic is affected by outdoor and indoor reliability ensuring that there is a discrepancy in latency and call setup times as well as cell-edge data rates that may affect the unlimited reliability desired for public safety networks. An algorithm to dedicate the transmission of voice over the networks while supporting videos over the commercial networks may be the short-term solution adopted before achieving interoperability within the broadband networks.

6.4 Public Safety Standardization

Embracing new broadband based systems for public safety networks requires elaborate mechanisms aimed at ensuring that they can access the mobile phone networks securely and integrated in a way that allows them to get the best out of them. Generic 3GPP releases are targeted at the mobile consumer business and bringing public safety systems on board would require a change of focus to make them safer and inaccessible to commercial users. Mobile technology has been dynamic and evolves with the increased advancement of newer systems intended to improve the systems. Waves of evolution have followed the introduction of 3G with Releases 8 and 9 defining the LTE technology and Releases 10 and 11 introducing LTE-Advanced (Ragaleux et al. 2015, p. 2754). Public safety requirements were introduced in Releases 12 and 13 with further enhancements following thereafter. With 5G and the internet of things becoming a reality, public safety functionality would have to be defined allowing safer integration of all the processes (Naqvi et al. 2018, p. 36).

6.5 Flawless Mobile Broadband for Public Safety and Security

Where coordination is crucial, wireless communication technology has emerged as being critical in coordinating different operational scenarios owing to the easy deployment over terrains where fixed technology is limited. Mobile communications and the internet have revolutionized different aspects of life. These systems work together in close associations with smartphones that produce performance with higher magnitude as compared with the first personal computers that were ever made. The current or existing computing and digital technologies have experienced huge progress in terms of data speeds, larger bandwidth, seamless mobility, and the Internet of Things. These features and characteristics are very effective to mobile networks because of their nature of enhancing flexibility, security, and scalability (Chavez et al. 2015). The commercial utilization of modern technologies has made the majority of wireless communications faster and effective. Wireless communication technologies have been very significant in regards to public safety such as in operational activities and timed communications. Support for mobility is detrimental to public safety networks.

A modern evolution of public safety, therefore, has been structured around mobile technology, creating the need for applications that would take advantage of the ability to transmit voice as well as reliance on data transmission. The efficiency of the disaster recovery operations, being one of the critical services, stands to benefit a lot from the integration of data-intensive multimedia applications. Critical information that enhances awareness would include video on location, downloading or uploading operational information to control rooms as well as online data inquiry on different aspects. Currently, dominant wireless communication standards like TETRA, APCO-25, and TETRAPOL provide secure voice services and data connectivity usable in emergency situations. They do not, however, offer broadband connectivity that would support applications. Some European countries have provided TETRA Enhanced Data Service (TEDS) for public safety together with wideband data connectivity enhancing the experience of critical services while providing a backbone for the development of data-intensive applications. Although TEDS is a step in the right direction, it should be noted that the gap between commercial networks and those used by the community public safety organizations remains big and governments need to invest in the public safety communications infrastructure. Lately, a radio frequency spectrum has been allocated to public safety broadband, expanding the range of available infrastructure that would augment the critical care services available through wireless technology (Tata and Kadoch 2014, p. 11). This chapter summarizes the potential applications available for public safety as well as the need to harmonize operations enabling the freeing up of resources that would otherwise hamper peaceful deployment of the networks for the benefit of public safety.

6.6 Applications in Different Scenarios

Public safety organizations support the variety of tasks enhancing the operational efficiency according to the demands of the community. Organizations at times require fulfilling different roles and having the right infrastructure eases the model allows for the efficiency of the processes. Critical care and public safety organizations are interrelated and often require accessing each other's resources to accomplish desired goals (Baskaran et al. 2017, p. 21977). Law enforcement units, as well as emergency units, for example, cannot be divorced from each other as they perform complementary roles in the society. Border security on the other hand equally relies on the efficiency of police services as well as the search and rescue units to accomplish its mandate. The inter-reliance on each other validates there should be ways of opening up the infrastructure in a way that would allow the quick interchange of ideas and approaches. The public safety operations operate in the different categories of the urban environment, which is synonymous with congested living spaces and tall buildings that may block the line of sight for wireless communication systems, necessitating the presence of alternative communication networks.

Rural environments have wide unoccupied areas that may require costly installations of some technologies to achieve universal coverage. Accessing the different networks complementing each other achieves maximum coverage. Radiofrequency networks would work brilliantly in such areas to complement the available land based communication technologies as well as wireless networks.

Critical care facilities like ports and airports share the same features with the urban environment but with the additional burden of carrying services that would rather remain operational even if there were interruptions in one form of transmission or another (Ahmad et al. 2017, p. 7989). Air traffic controllers, for example, operate extremely critical service that may result in loss of life if it were interrupted without a complementary network in place. Equally, its importance in the maintenance of safety validates the essence of maintaining the most stable and functional networks.

Efforts to bring on board public safety and security systems are not just aimed at addressing voice communication systems and the ability to send short message systems but to reimage the whole concept as well. Changing the systems as they are would completely change the conception of public safety since the networks would be shared but robustness put in place to address safety concerns. Public safety organizations have developed a robust voice communications network that uses Land Mobile Radio (LMR) while others use Terrestrial Trunked Radio (TETRA). The conventional best effort basis infrastructure operates the data communication networks which do not, however, integrate with mission-critical communication. Disparities between the approaches represent a major shortcoming in terms of public safety networks. Commercial networks may represent the future in terms of communication and data transmission but there are insufficient safety protocols for the development of apps that would guarantee the safety of mission-critical communications (El Alaoui 2012, p. 37).

Harnessing the gains made by the cellular communication networks for commercial users and integrating them for use by the public safety technology networks should be the goal for the development of new safety standards. Current mobile technology based on 4G and 5G networks have become, for example, robust enough to have the capacity to recognize facial features in individuals, even in a group photo. While the technology is used for accessing different devices and analyzing social media posts, there are enormous benefits that mission-critical communications would derive from such technology in offering their services. Offering real-time facial recognition technology would, for example, be of profound benefit in instances where the mission-critical service providers require recognizing suspects in disasters or as suspects for apprehension. Technology is developed at the moment but has not been enhanced with the features that would allow real-time deployment on dedicated data traffic accessible by mission-critical service providers to enhance public safety (El-Keyi et al. 2017, p. 20813). Marking the desired faces from crowds in public events would additionally help in mapping out the bad elements in society and maintain rock-solid surveillance, especially in real time. First Responder Network (FirstNet) and the Next-Generation Communications have the responsibility of providing resilient broadband that enhances life-saving missions for first responders. Lack of broadband for first responders makes personnel carry their own smartphone devices that are typically used by members of society but without any added benefit of a dedicated infrastructure for analysis as part of community safety measures. Adopting FirstNet networks for use by mission-critical services would allow for the development of resilience, especially in times of natural disaster when the public infrastructure may be compromised.

6.7 Public Safety Systems and Architectures

The modernization of networks, applications, broadband internet, and social media have become opportunities and challenges for stakeholders in public safety. The Department of Homeland Security in the United States has worked with municipalities to update the National Public Safety and Emergency Communications plan. In their plan, building infrastructure and networks for voice, video, and high data rate are essential services to communicate and share information in all instances. The following is a description of current and future network infrastructure for public safety and emergency communications systems.

6.7.1 Airwave

This is a British mobile communications company that is also the largest private operator in public safety networks. It offers mission critical voice and data communications to over 300 emergency agencies in Great Britain. Its main technology revolves around Terrestrial Trunked Radio (TETRA) technology provided by Motorola solutions. One main objective for this organization is to ensure the establishment of a secure, resilient, and interoperable network. However, in recent developments, the Emergency Services Mobile communications Programme (ESMCP) is set to fully replace TETRA communications using LTE technology.

6.7.2 LMR

As stated earlier in this chapter, also known as PMR communications, public safety networks in the recent past have experienced advancement toward broadband. For example, Terrestrial Trunked Radio (TETRA), as one of the earliest systems in these networks, transformed from narrowband to TEDS wideband (TCCA 2018). TEDS is capable of offering higher data rates that go up to several hundred kbps, making it one of the effective networks within the PMR landscape. In Northern America, the APCO25 has experienced similar transformation.

Ideally, public safety LMR systems provide mission critical communications for first responders and are considered essential to managing day-to-day agency operations and response to emergency incidents. First Responder Network (FirstNet) has estimated that there are more than 60 000 public safety agencies in the US providing law enforcement, fire, and emergency medical services (EMS). All of these agencies are served by some form of public safety LMR system that includes push-to-talk (PTT) voice communications.

Some public safety agencies have implemented PTT services provided by commercial cellular providers, allowing them to shift administrative and support personnel to a non-mission critical communications system and lower operational costs. This service is called "push-to-talk over Cellular" or POC. The Nationwide Public Safety Broadband Network (NPSBN) being implemented by FirstNet will be the first fully interoperable network supporting data, voice, and video for all first responders in the US. The NPSBN is envisioned to transform public safety operations through the provision of new services, features, and capabilities. According to Paulson and Schwengler (2013) one of the proposed new services is mission-critical push-to-talk (MCPTT), which is being designed to provide some LMR-like services that may eventually allow first responders to carry a single device to access voice, video, and data. However, comparability with existing LMR systems for voice is yet to be determined, (e.g. for coverage and direct unit to unit communications).

Public safety agencies rely on mission-critical voice communications to conduct daily operations and to manage complex emergency events. Public safety LMR systems are purposely built and designed to support individual and group communications. These systems are typically voice centric and operate as a self-contained local network. Public safety LMR systems also provide crucial off-network service allowing direct communications between first responders without the need for an infrastructure. There are a wide range of LMR systems in use by public safety agencies today. These include analog and digital systems, networks that use simplex channels, conventional repeaters, and trunking radio technology. Three forms of LMR have defined standards in North America. They include P25 trunking, P25 conventional, and analog. Some systems are built using proprietary digital solutions offered by a specific manufacturer. The type of system built by a public safety agency is based on a number of factors including cost, coverage, features, and need for interoperability. Public safety LMR networks operate in many frequency bands including very high frequency (VHF), ultra-high frequency (UHF), 700, and 800 MHz. Additionally, public safety agencies use a variety of console and gateway systems to operate and manage their LMR networks. This use of disparate spectrum and different technology creates interoperability challenges (Paulson and Schwengler 2013).

The adoption of the P25 digital radio standard has greatly enhanced the interoperability landscape. There are a number of solutions in use today to create interoperability between public safety LMR systems. These include simple console patching and the use of IP gateways as well as other mechanisms to support local needs. Some of these interoperability solutions will likely be leveraged to support LMR–LTE interoperability (Frosali et al. 2015)

6.7.3 TETRA Security Analysis

Terrestrial Trunked Radio refers to a European Telecommunications Standards Institute standard, which came to force in 1995 for the first time. The design of this standard targeted mobile communication systems, whose primary design was targeting security personnel for use.

Apart from the law enforcement units, it targeted emergency, rescue and service operations, and the people involved in public transportation. In general, this feature would support the national safety communications network (Oueis et al. 2017). Terrestrial Trunked Radio is built and is currently in operation in over 100 countries around the world. The main service of Terrestrial Trunked Radio is the voice communications feature. Terrestrial Trunked Radio has some special features, which include:

- PTT group-calling feature/mode
- Direct terminal–terminal radio transmission
- Short call setup time feature.

Terrestrial Trunked Radio provides users with authentication protocols, which consist of:

- Radio channel
- End-to-end encryption feature.

In this report, there is going to be a description of the Terrestrial Trunked Radio security system, which includes the following:

- The cryptographic primitives Terrestrial Trunked Radio uses
- The encryption protocols and authentication Terrestrial Trunked Radio uses
- Terrestrial Trunked Radio key management
- The mechanisms and key management given for cooperation across Terrestrial Trunked Radio security domains.

This will basically be a security analysis of the system, which will take the shape of setting up security assumptions, which will envision the attacker model and then analytically describe the result of the analysis, which will expose the merits or demerits of this feature, whichever are stronger (Oueis et al. 2017).

This is, therefore, a formal security analysis of the protocols of authentication as they could happen. Eventually, the information from this analysis will attempt to bring out a descriptive understanding of the possible technical problems of Terrestrial Trunked Radio if any exist in connection with the communication security Terrestrial Trunked Radio.

6.7.4   TETRA Services System

Trunking refers to the technique deployed in radio systems in order to have the availability of communication expanded to have a multiplicity of the resources. For example, trucking facilitates universal and widespread access to all users, such that they will have access to all the channels automatically. There are two assumptions of this system:

1) The average message to disseminate will be short and therefore several stations need to have simultaneous communication.
2) There are three types of trunking, which include

- Transmission trunking
- Message trunking
- Quasi-transmission trunking.

In the message trunking service, the system assigns a radio channel for the whole period of conversation. In the transmission trunking, the system assigns a radio channel for a short duration, equal to a radio transmission in a single half-duplex format.

Owing to the need for every transmission to obtain a channel, during busy hours subscribers might experience delays. In the quasi-transmission trunked systems, however, the requests from recently terminated groups will receive high priority over other requests. The mechanism would, therefore, guarantee the continuity of a message for end users (Oueis et al. 2017).

TETRA refers to a standard developed for environments of private mobile radios. The frequency for its operations is approximately 150 MHz, and it has the capability of offering a bit rate that goes up to 28.8 kbps.

6.7.5   The Architecture of TETRA

The TERTA network has various interfaces. In this section, the discussion takes a brief look at the architecture of the TETRA network. Figure 6.1 gives a general overview of the architecture of the TETRA network.

6.7.5.1   The Interfaces of TETRA Network

The TETRA network has various interfaces numbered below and outlined in Figure 6.1:

- Interface 1 is the radio air interface
- Interface 2 is the line station interface
- Interface 3 is the intersystem interface. Interface 3 permits the interconnection of the TETRA networks by different manufacturers.
- Interface 4 refers to the terminal equipment interface for the mobile station, which also refers to the terminal equipment interface for a line station
- Interface 5 refers to the network management interface
- Lastly, interface 6 refers to the direct mode interface.

6.7.6   TETRA Network Components

The network components refer to the functional structure of the TETRA Network. They include the following:

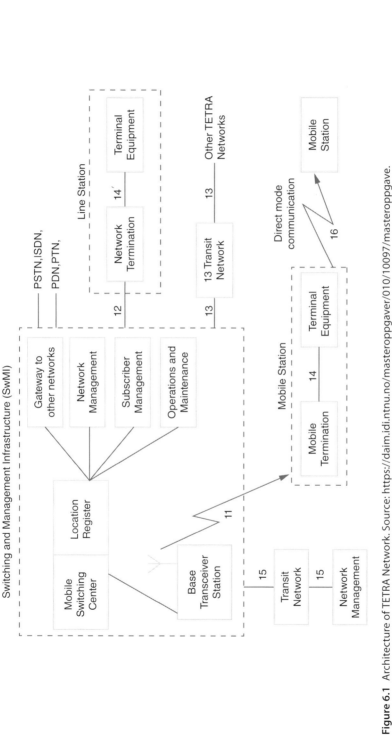

Figure 6.1 Architecture of TETRA Network. Source: https://daim.idi.ntnu.no/masteroppgaver/010/10097/masteroppgave.

6.7.6.1 The Mobile Station

The mobile station has components that include the physical equipment of the subscribers. There is also an identity module known as the subscriber identity module, and lastly, there is the TETRA equipment identity, which is specific to every mobile device. The TETRA equipment input is a system that allows the operator to input the equipment identity. In this system, it is possible for the operator to disable an equipment deemed or reported stolen immediately.

6.7.6.2 TETRA Line Station

The TETRA line station operates in a similar structure and format to the operations of a mobile station. In this system of operation, it is able to switch and manage the infrastructure it has connections with, over a integrated services digital network (ISDN). In this interface, it gives the same functionality services to those given by the mobile station (Guoru et al. 2018).

6.7.6.3 The Switching Management Infrastructure

The switching and management infrastructure is a type that contains the base stations that are responsible for the maintenance and the establishment of the core communications happening between the line stations and the mobile stations over the ISDN. The switching and management infrastructure allocates switches calls, and channels, and has databases that contain the subscriber's information station (Guoru et al. 2018).

6.7.6.4 Network Management Unit

This interface gives the remote and local management functionalities.

6.7.6.5 The Gateways

The gateways have the function of interconnecting the TETRA network and the non-TETRA network like the ISDN, the packet data network (PDN), and the public switched telephone network (PSTN). The conversion or translation of the information formats and the protocols of communication may be necessary for this protocol.

6.7.6.6 How the TETRA System Operates

The TETRA system has identities that are useful in distinguishing the communication parties in the TETRA network. During the manufacture of the equipment to use in the TETRA network, the equipment at the stage of manufacturing gets a TETRA equipment identity, which is akin to the IMEI of phones or the international mobile equipment identity that the GSM technology applies. In the TETRA equipment identity system, the contents are as in the illustration in Figure 6.2.

Here the mobile network identity is responsible for identifying the different forms of TETRA network. The bases station after this identification broadcasts the mobile

Type Approval Code (TAC)	Final Assembly Code (FAC)	Electronic Serial Number (ESN)	Spare (SPR)
24 bits	8 bits	24 bits	4 bits

Figure 6.2 The TEI/TETRA equipment identity.

Figure 6.3 TETRA subscriber identity.
Source: https://daim.idi.ntnu.no/
masteroppgaver/010/10097/
masteroppgave.

Mobile Country Code (MCC)	Mobile Network Code (MNC)	Short Subscriber Identity (SSI)
10 bits	14 bits	24 bits

network identity. The mobile network identity works in a way such that it broadcasts the operator information as well as the country code or any of the two.

The TETRA subscriber identity is useful in relating the subscribers with their billing information and services subscribed. The TETRA subscriber identity has the contents shown in Figure 6.3. TETRA has a unique service system, which is the group-call service system. Apart from the individual TETRA subscriber identity, the TETRA service has a unique feature, which is the group TETRA subscriber identity.

6.7.7 TETRA Mobility Management

TETRA has a mobility management system that works on similar terms and mobility system to the GSM. It has what is known as the home database, which is the holding point of information of the mobility management system such as the subscribed services by the users, the cipher keys, and the identities of the users. Authentication occurs in the visited networks through the home database and the essential user information downloads to the visitor database feature station (Guoru et al. 2018).

6.7.8 The Security of TETRA Networks

In this section, we address the important security features of the TETRA network. Here the scope of coverage spans the authentication, encryption mechanisms, and the key management of TETRA. It dictates that a secure communication network needs to provide (Stavroulakis, 2007):

- Confidentiality
- Integrity
- Reliability
- Non-repudiation
- Authentication.

6.7.8.1 Confidentiality
Only authorized personnel or people should have access to the information being passed along.

6.7.8.2 Integrity
This refers to the requirement that states that only authorized users need to be able to make any modifications to the information in exchange.

6.7.8.3 Reliability
This refers to the requirement that the resources and the services are not denied and are available to the authorized users to accomplish various tasks.

6.7.8.4 Non-repudiation
This requires that the sender cannot deny that he/she sent the message.

6.7.8.5 Authentication

This refers to the requirement that the sender's identity is verifiable by the recipient.

6.7.9 The Process of Authentication in TETRA

In TETRA, authentication, like in the other systems of communication, refers to a fundamental security service that safeguards the users of TETRA. Authentication is, therefore, a process that ensures that the parties who participate in the communication process prove their identities. In the event there is the use of public resources the proof of one's identity must be through a signed certificate issued by an authority.

In the event of using a symmetric key cryptography, there is an assumption of a high level of security hence one can share the same secrets with another person. It is only through the secret or code that people can communicate. The basis of authentication in TETRA is proving that one knows the secret shared between two parties, for instance the authentication center and the mobile station.

6.7.10 The Authentication Key

The authentication key uses some basic principles and processes. These are cryptography classes of algorithms.

There are two classes in this system for cryptography, which are

- The asymmetric key algorithms
- The symmetric key algorithms.

6.7.11 Symmetric Key Algorithms

In the symmetric key algorithms:

- The sender and the recipient share a similar or the same secret key.
- When people apply the symmetric key algorithms and use systems containing multiple nodes, the leakage of the secret key at any of the given nodes will compromise the entire system as it will make the entire system insecure.
- It is therefore important to take preventive action, which is the periodic update of the secret code/key.

Block ciphers and stream ciphers:

- Symmetric key algorithms have steam and block cipher schemes.
- Block cipher schemes actively divide the plain text into fixed length blocks and it encrypts the block systematically.
- A stream cipher, on the other hand, operates on the principle of one plain text and one digit at a time.
- In this mechanism, a secret key helps in the initiation of a pseudo-random keystream sequence, which plays the role of combining with the plain text.

Public key/asymmetric key:

- In this key algorithm, the decryption and the encryption processes utilize different keys.

- In the asymmetric key, the cryptography uses a mathematical function in place of the substitution function or permutation.
- This process is computationally unfeasible especially when one wants to extract one of the keys from the other key and the cryptographic algorithm.
- Every user here possesses a pair of keys, which are in the form of public and private keys.
- The public key is public while the private key is private.
- The moment two parties attempt to communicate, the sender needs to know the public key of the receiver in order to encrypt the message using the recipient's public key.
- When the recipient gets the message, they can use the private key for decrypting the message.

6.7.12 The Process of Authentication Key Generation

In TETRA, the process of authentication key generation utilizes symmetric keys, see Figure 6.4. Here, the mobile station is assigned a user authentication key when it is registering to the network for the first time (Adibi et al. 2008).

The user authentication key is thereafter stored in the SIM card of the terminal equipment as well as stored in the database of the authentication center. K denotes the authentication key, which is the knowledge that must be demonstrated in order for authentication to take place. It could be generated in three different ways. These include the following:

- From the AC or the authentication code, which is the pin code that the user enters.
- Generated through the TB1 algorithm from the user authentication code stored in the GSM, which has the algorithm TB2.
- Utilizing both the user account code (UAC) and the AC for generation of K is the third method. In Figure 6.4 it is TB3.

It is important to note that the lengths of the KS', K, and KS are all 128 bits. K cannot be subject to direct use in the process of authentication. However, it can be usable in the generation of the session keys, which are KS' and KS.

6.7.12.1 ESN (In United Kingdom)

The Emergency Services Network (ESN) was set up by the government to replace the airwave service. According to the United Kingdom's government, in future, ESN has

Figure 6.4 Authentication keys generation. Source: http://daim.idi.ntnu .no/masteroppgaver/010/10097/ masteroppgave.

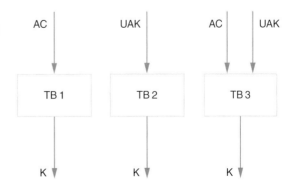

the potential of offering "next generation integrated critical voice and broadband data services as a mobile communications network with extensive coverage, high resilience, appropriate security, and public safety functionality" (Kable 2016). This organization, just like the European Emergency Number Association (EENA) covers three fundamental emergency services, fire, police, and ambulance.

6.8 Emergency Services Network (ESN) in the United Kingdom

The United Kingdom recently began implementation of the Emergency Services Network (ESN) to replace the Airwave Communications Network, which has been in use for quite some time by the fire and rescue, police and ambulance services in the United Kingdom.

6.8.1 Overview of the ESN

The ESN is a project of the United Kingdom Home Office as a part of an upgrade to a new critical communication system delivery. It replaced the Airwave Service that the emergency services used in Britain (Adibi et al. 2008).

ESN customers include the fire and ambulance services, the rescue services, police, and other users like local authorities and the providers of utility services like first responders in the water bodies such as inshore rescuers.

The ESN service serves over 300 000 people in the United Kingdom using handheld devices, as well as operational equipment fitted in vehicles, aircraft, and control rooms across the United Kingdom. The ESN runs on software provided by Motorola Solutions. It operates on 4G radio frequencies, which are usable both in urban and rural areas in the United Kingdom. ESN has its focus on making the services required by the clients fast at the time the clients need the services. This implies that the main elements of ESN are testable, adaptable, and usable on demand (Adibi et al. 2008).

6.8.2 The Deliverables of ESN

The United Kingdom began implementing ESN with the strategic aim of delivering better services and voice data to the emergency services. It was going to replace the limited yet reliable Airwave system to transform the operations of emergency services mobile operations, especially in remote areas, when there is congestion in the networks. The aim of the ESN is to create a single platform for the sharing of imagery and data to enhance the faster adoption of successful mobile applications. The ESN is representative of the taxpayer's value for money as it delivers steady-state savings, which account for UK£200 million annually (Adibi et al. 2008).

6.8.3 The Main Deliverables of ESN

- Resilient and secure communications critical for emergency services missions as a more trusted way to keep the workers of these services safe.
- Provision of a mobile data platform, which enables the emergency services to improve their frontline operations and have more closely coordinated work.

- Provision of increased value for money as it replaces a legacy technology, and improves the commercial terms as it provides a common platform for conducting innovation and sharing of data.

6.9 SafeNet in South Korea

In South Korea, the motivation for the creation of the SafeNet project came from the lack of an agency that could facilitate interoperability, which could have helped in responding to several disasters that occurred in the recent past.

SafeNet is the ambitious creation of South Korea that aims to create a public safety network on the commercial LTE infrastructure. This creation of South Korea was an ambitious project, which South Korea rolled out by 2017. The main center of this system in Seoul, where there is the main control center and a center that hosts the core equipment that runs the system (Oueis et al. 2017).

South Korea pioneered the system in early 2016 in the province of Gangwon as it transitioned to the SafeNet system. Before the SafeNet program delivery, South Korea was using a patchwork of separate LMR networks to deliver the internet.

At the time of rolling out the SafeNet program, the priority of South Korea was rolling out the system in the country's rural districts, which already had the legacy local LMR networks. This marked a major revolutionary upgrade for the rural areas of South Korea.

The country awarded the contract to the second largest MNO company in South Korea. The company that won the contract had the task of providing capacity and coverage to support 200 000 users who came from 324 national agencies and partnering organizations throughout South Korea (Oueis et al. 2017).

Unlike the network projects in Queensland and Los Angeles, SafeNet delivers voice and data over LTE. In this delivery, there are PTT applications for multimedia along with critical voice communication functionality.

While the technologically is similar to the ESN solution, SafeNet utilizes a dedicated spectrum instead of sharing a spectrum with a wider user base. The Ministry of Public Safety and Security manages SafeNet in-house in collaboration with the Third Generation Partnership Project also known as 3GPP.

Samsung is the main contractor that delivers the end-to-end services, which include additional mobile cells, networking equipment, and rugged smartphones. This contract includes the core hardware supporting the data-rich infrastructure, equipment, and PTT voice. Additionally, it includes an international standards body release, which introduced a criterion called MCPTT. This criterion is the first for safety agency users in Gangwon and Seoul, who were the first to use such capabilities on a large scale of 3.4.1, which offers opportunities and strengths (Oueis et al. 2017).

Through working in concert with the industry, SafeNet is able to work with other partners in order to create resilience of the system.

Apart from South Korea's heavy industrial base, it has a large workforce that is associated with deep sea fishing. This deep sea fishing feature covers the South Korean maritime and waterways. The system is helpful to South Korea because it is helpful in facilitating quick responses to maritime incidents.

An incident in South Korea pushed the country to pursue its first national inter-agency public safety network. The sinking of MV Sewol made the government of South Korea see the sense in creating such an integrated system. One monumental aspect of this integrated system connects with South Korea's largest mobile network operator. Despite losing out on the contract to create the system, SK Telecom, which is the biggest network operator in South Korea, has come up with LTE high-speed data communications that offer ships communications data as far as 100 km offshore.

The technology of KT telecoms aligns to the SafeNet and other 15 first responder solutions used in the United Kingdom, thus offering an interoperable network. While South Korea is a leader in LTE coverage, it has low coverage of incidences. The reason for this is that the South Korean MNO market experiences a highly oversaturated and competitive condition (Kumbhar et al. 2017).

South Korea in this sense can achieve a high sense of interoperability, especially in its heavily technological and industrially intensive environment. South Korea can, therefore, benefit a lot from SafeNet as it can source the best breed of commercial solutions, and it can also access the best market innovations.

For instance, KT has developed a private LTE network service that is an enterprise service that works in a remarkably similar fashion to the way SafeNet works. With the current stagnation experienced in the commercial market, the suppliers of SafeNet as well as other interested parties know that the only way they can increase their streams of revenue is through the sales of similar systems to clients outside South Korea. Therefore, these people have the drive to ensure the success of the network (Kumbhar et al. 2017).

Although there are weaknesses, threats and lack of coverage, and the potential for disastrous outcomes, SafeNet does not have these as key concerns because these risks are largely under the mitigation of very high commercial provision.

The main worry, however, is the network capacity in specific critical PTT voice applications that command higher bandwidths. Innovative backup solutions are under progressive design phases in order to have the coverage and supply capacities expanded, especially in the areas of low connection.

One of the significant backup solutions, which doubles up as an innovative step is the design that is ongoing seeking to expand the supply capacity and the coverage in the low connection areas. This is inclusive of the base station, which has a backpack unit, and an unmanned aerial vehicle, which has camera equipment and networking modules, which are important for the maintenance of connectivity. The effectiveness of this equipment, however, depends on the staff (Kumbhar et al. 2017).

The rollout of the networks has its priority in the rural areas over the urban areas, which already have LMR networks. This step is very important because it is strategic in the sense that it seeks to ensure both urban and rural areas have active LMR networks that can enhance rapid service and security provisions.

The rollout program relies heavily on the capabilities of staff being able to establish the backhaul infrastructure, which other users can also share, especially staff who have little experience in working collaboratively. Working in collaboration with the 3GPP implies that SafeNet is able to benefit from the commercial standards, which results in overall cheaper costs than a more specialized solution that can be found. This also implies that the network may fail to be optimized for public safety uses.

Whilst having a partnership with 3GPP, SafeNet is under the management of the Ministry of Public Safety and Security. One of the main challenges of SafeNet in South Korea is the issue of interoperability when responding to disasters. As per the goals of South Korea, by 2017 the country was determined to swiftly create a public safety network over the commercial LTE infrastructure. Before the establishment of SafeNet, users heavily relied on separate LMR networks. For SafeNet in South Korea, it is meant to roll out in rural districts as a majority of the Metropolitan areas are characterized by local LMR networks.

Toward the end of 2017, Samsung was contracted to offer end-to-end services such as extra networking equipment, mobile cells, and smartphones. This was inclusive of the core hardware offering support to the PTT and data rich infrastructure tools. Moreover, as the international standard body, 3GPP created the MCPTT criteria to be utilized by public safety agencies and users in Seoul and Gangwon (Kable 2016). One of the strengths of SafeNet is its interoperability with SK Telecom, which is the largest mobile network operator delivering LTE high speed data communication.

6.10   FirstNet (in USA)

FirstNet was created in 2012 via congressional legislation to build an NPSBN. A major component of the NPSBN is to enable and support interoperability. Indeed, the genesis of FirstNet came from the 9/11 Commission Report, which noted that a lack of voice interoperability disrupted emergency operations and was life threatening for first responders (Kable 2016). In order to fully support first responder interoperability, Congress required that FirstNet build the network using open, non-proprietary standards (First Responder Network Authority 2012). The adoption of open standards is an essential component to meet the need for interoperability between LMR and LTE networks and is necessary to adopt the requirements and recommendations contained in this report by the National Public Safety Telecommunications Council (2018). FirstNet is also the First Responder's Network Authority of the United States of America. The establishment of FirstNet came as a part of the tax relief program and a job creation act, which was a nationwide creation of a Public Safety Broadband Network. This is a preserve of the public. FirstNet is a nationwide establishment in the United States, which creates an interoperable public safety broadband network, which is a preserve of the first responders in the United States of America. In the establishment of the network, FirstNet follows some guidelines or principles. There are several principles guiding the creation of FirstNet.

1) FirstNet: Meeting the United States First Responder Needs
   Firstnet has to work with the following agencies to meet the needs of first responders:
   - The National Public Safety Telecommunications Council (NPSTC)
   - The Association of Public-Safety Communications Officials (APCO)
   - The FirstNet Public Safety Advisory Committee (PSAC).
2) FirstNet Shall Provide Public Safety Users with Actual Priority Access to the Network
   - The core reason for the building of FirstNet is public safety in the United States
   - The network has the purpose of providing broadband wireless communications to law enforcers, paramedics, firefighters, and other public support and safety personnel to meet the important daily missions of these agencies.
3) Role of Hardening the Network to Help with Resilience in Times of Incidents, Natural Disasters, and Threats Instigated by People
   - Hardening here comprises strengthening of the overall network and cell tower sites to maximize communication reliability

- Construction and designing of a network that is both redundant and resilient
- Engineering a network with backup equipment and backup services for sustaining operations in times of adversity
- Hardening the communication infrastructure to withstand incidences like flooding, earthquakes, and winds
- Creation of guidelines for all the radio access network components
- Hardening the power supplies, antennas, towers, towers, temperature controls, electrical, and physical connections going to the user devices from the network
- Determining the best ways of addressing the hardening required for data centers, servers, and points of aggregation.

4) Enhancement of public safety communications through the delivery of mission-critical data and applications for augmenting the voice capabilities of the LMR networks
   - The functionalities here entail the provision of high-speed data services for supplementing the LMR networks voice capabilities
   - Providing users with capabilities for sending and receiving data in forms of images, data, video, voice as well as text
   - Enhancing user communication capabilities of communicating over the network and benefiting the users by providing abilities for sharing applications.

5) FirstNet to Enable the Management of Local Communications and Keep Incident Commanders in Control
   - FirstNet has a responsibility of enabling and local communication to deal with local incidences locally
   - Creation of parallel hierarchical avenues for incident management and communication
   - Providing a single safe nationwide safety broadband network, which has dimensions for local control for addressing the local incidence
   - Empowering the agencies at the local level to determine the users with priority for using the network to ensure satisfaction of the public safety priorities

6) Role of Using the Taxpayers Dollars Judiciously While Keeping the Focus on Offering the Services to Public Safety
   - Delivery of valuable services and applications as well as networks tailored to the public community safety requirements
   - Ensuring creation of structures to serve the unserved areas
   - Enabling multiple jurisdictions to share the access to applications cost effectively through information sharing on common database applications like criminal background information and motor vehicle particulars
   - Ensure that it is self-sustainable by collecting only the fees necessary to recoup their expenses by creating a suitable pricing model to attract paying users
   - Offering friendly pricing models to enable all public safety users to be able to access and benefit from everything that the network has to offer.

7) FirstNet must have effective Security Controls that Protect Data and Create Defense Against Cyber Threats
   - FirstNet shall have effective security controls, which defend against cyber threats and protect information
   - Offer effective defense against the rapidly changing and complex security threats
   - Building layers of security at every determined vulnerable point

- Designing security into all the radio access networks
- Designing security into the evolved packet core networks
- Designing security into the service platforms
- Designing security into the devices, which use the built network
- Creation of firewalls that facilitate the enforcement of stringent security policies developed with cooperation with the following departments:
  - Department of Homeland Security
  - Department of Defense
  - Developed to satisfy the requirements of National Institute of Standards and Technology (NIST)
- FirstNet will use the guidelines of 3GPP in the encryption as well as other best-determined security measures that follow the best practices
- Working closely with Federal agencies that have expertise in the security design modeling of telecommunications.

6.10.1 The Benefits of FirstNet

FirstNet is helpful in several ways ranging from large events and emergencies to day-to-day operations. People's lives depend on the ability of the public safety authorities to communicate. FirstNet in this respect provides reliable connections, which makes it possible for first responders to send and receive data confidently. The data that FirstNet safely disseminates include voice data and text data transmissions, with the lifesaving mission demand securities.

FirstNet gives its subscribers the opportunity to transformthe emergency communications, which they have through secure wireless connectivity, which permits the agencies to achieve mobile data integration into their core operations.

6.10.2 Public Safety Core of SafetyNet

SafetyNet has been built specific to the needs of the First Responder Network Authority, developed through several years of consultation with the American first responders in all territories and states and on all the levels of government (Dunlop et al. 2013).

As such, the FirstNet core is the network foundation, which ensures the delivery of safety features for advanced public safety, which is unique to FirstNet. These features, which also represent the benefits of FirstNet include the following

6.10.2.1 End-to-End Encryption
In end-to-end encryption, FirstNet has the following advantages, which come in the form of superior security features:

- FirstNet core has Federal Information Processing Standard (FIPS) 140-2 amenable VPN solutions
- Radio encryption
- Transport encryption
- Network core encryptions
- Advanced logical and physical security protocols, which help in keeping all the network and traffic under stringent protective measures.

6.10.3 Round the Clock Security Surveillance

Under the round the clock security surveillance, FirstNet has several advantages. These advantages include:

- A system of constant monitoring by a security operations center using dedicated and state of the art tools as well as dedicated security experts working for FirstNet.
- The security team of FirstNet has the sole duty of monitoring the network security and protecting the important data, services, and applications running on the FirstNet network.

6.10.4 User Authentication

FirstNet employs a system of identity credentials and access management (ICAM) solutions whose focus is to create simplicity in the system and create ease of use to the people while ensuring that there is a positive identification on the system for both the users and the devices.

6.10.5 Mission Critical Functionalities

The FirstNet core has an efficient system that delivers a wide range of futuristic public safety capabilities, which are currently under development based upon open standards and are specifically for the safety of the public.

These services will in the future include services like the MCPTT and enhanced services like location-specific services.

6.10.5.1 Tactical LTE Coverage

Throughout the discussion, it has been established that LTE and 5G network communications are the future of public safety networks. In a more comprehensive way, two authors named Ferrús and Sallent (2015a,b,c) came up with a table showing a summary of the differences existing between commercial network operators and public safety network systems in a report titled *LTE Networks for PPDR Communications. Mobile Broadband Communications for Public Safety: The Road Ahead Through LTE Technology.* Furthermore, Table 6.1 shows the suggestion the two authors tabled regarding improving the performance and functionality of public safety networks and the general future of these communication technologies.

6.11 Canadian Interoperability Technology Interest Group (CITIG)

The Canadian Interoperability Technology Interest Group (CITIG) is an organization that encompasses representatives from public safety, industry, academia, and government to be in charge of all the activities related to interoperability between Canadian public safety providers. It was created in 2007 by the Canadian Police Research center (CPRC) and the Informatics Committee of the Canadian Association of the Chiefs of Police (CACP). Since it is an organization dealing with issues of public safety, its objectives include:

Table 6.1 Commercial network operator versus public safety network model.

Issues	Commercial network operator model	Public safety network model
Goals	Maximize revenue and profit	Offer protection to life, property and state
Capacity	Often defined by a "busy hour"	Often defined by "worst case scenario"
Coverage	Population density	It is territorial, focuses on whatever may need protection
Availability	Outages are undesirable	Outages are unacceptable because lives may be lost
Communications	One-to-one	Encompasses dynamic groups
Broadband data traffic	Internet access mainly downloads	More uploads as compared to downloads
Subscriber information	Ownership is identified to a carrier	Ownership is by an organization
Prioritization	Often has minimal differentiation based on the subscription level or the application	There is significant differentiation by role or the incident level
Authentication	This is carrier controlled, and device authentication only	It is organizational controlled with user authentication
Preferred charging method	There are subscriptions with predefined numbers of minutes, for example, per minute for voice, per GB for data, per message for SMS, all these with a fixed price	There can be quarterly or annual subscriptions with limited use

- Addressing challenges discovered while conducting research for interoperability
- Encouraging economies of scale and establishing a more standardized approach to interoperability across the country
- Providing a unique and effective information sharing platform to be of benefit to the public safety community
- To limit the obstacles facing interoperability solutions to life.

6.12 Centre for Disaster Management and Public Safety (CDMPS) at the University of Melbourne

Just as its name suggests, this organization in Melbourne was established to create a richer and smarter platform to support timely and effective decision making in relation to disaster management. Since it is responsible for ensuring public safety, this organization recognizes that disaster events are always increasing in frequency and severity. According to them, the trends associated with disaster management include the notion that disasters are evolving and expanding and there needs to be a policy driven change as well as the creation of strategic frameworks.

6.13 European Emergency Number Association (EENA)

EENA, which stands for European Emergency Number Association, is a non-governmental organization based in Brussels. It was established in 1999 with the aim of ensuring high-quality emergency services. Throughout the European Union countries, the number with which to seek emergency help is 112. Based on its organizational framework, EENA serves as a discussion platform for emergency services, decision-making, public authorities, and enhancing emergency responses as per the citizens' requirements. In addition, this organization also has the responsibility of warning the citizens prior to imminent disasters to ensure their safety. Their memberships encompass over 400 emergency services with representatives from 39 European countries. There are also 25 solution providers and 20 members in the European government. Figure 6.5 is a pictorial presentation of the services involved in the 112 emergency line.

There are five models present in the local public safety answering points (PSAPs). For instance, in model 1, the main characteristic is that all calls channeled to 112 are routed to a local emergency service. It is important to note that the person receiving the call belong to one discipline, which is comprised of fire, police, or EMS. Secondly, in model two, the main characteristic is that the call becomes filtered through thenational PSAP stage 1, and thereafter transferred to the regional services. The third model involves the collection of all disciplines in a single location at the regional level. In model 4, very experienced and well-trained call-takers in charge of 112 calls engage in calls and dispatch. Finally, in the fifth model, 112 PSAPs interconnect a single network using the same technology. For example, a call in region Y can be handled while in region X.

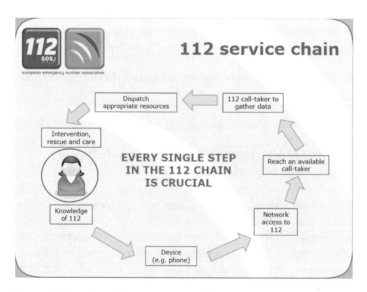

Figure 6.5 Services of the 112 emergency line.

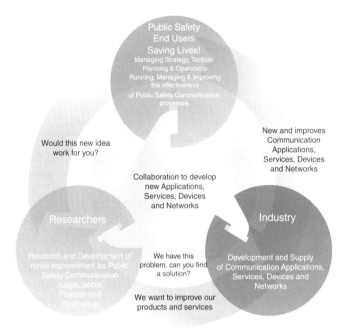

Figure 6.6 Services of Public Safety Communications – Europe (PSCE).

6.13.1 European Standardization Organization (ESO)

This is a regulation organization which implements the requirements, in this case mobile broadband data services and other ad hoc network services to various European nations.

6.13.2 Public Safety Communications – Europe (PSCE)

Figure 6.6 represents a summary of what Public Safety Communications –Europe (PSCE) is all about.

Briefly, PSCE is an organization dealing with the public safety communications with the aim of involving users and citizens, influencing EU stakeholders, raising awareness of the issues facing public safety through research, and informing the members regarding the new technological advancements, problems, solutions, and research outcomes.

6.13.3 The Critical Communications Association (TCCA)

TETRA and The Critical Communications Association (TCCA) is a section of a public safety association that is responsible for over 160 public and private users, as well as the manufacturers of mission critical communication services. By definition, mission critical communication refers to the specialized communication systems that comprises of availability, reliability, stability, and security of mobile communication that ensure there is continuous function of critical aspects in society. In the modern world, these kinds of communications are mostly known to be used by police departments and emergency services. It is also important to note that TETRA is responsible for the implementation of PPDR broadband services in the member states (Ferrús and Sallent 2015a,b,c).

This includes an array of activities such as creating a dedicated network or outsourcing to a commercial operator. The PPDR spectrum range must be in accordance with the spectrum regulator community (CEPT-ECC).

6.14 Public Safety Network from LTE to 5G

LTE networks are multi-purpose and data centric. They are capable of supporting different types of voice, data, and video services. This includes full-duplex voice communication (e.g. telephony and video chat) as well as PTT voice. Commercial cellular companies provide PTT via POC applications that may be designed for different user groups and are typically divided into two categories, PTT and MCPTT. While MCPTT is based on 3GPP international standards that are continuing to evolve, several vendors offer POC solutions today. These may be implemented as either a network service or an over-the-top application. While true MCPTT is not available today, some agencies are using existing POC solutions during incident response and in mission-critical environments. Just as interoperability between LMR and MCPTT is essential, so is interoperability between first responders who may be using different POC services (Balachandran et al. 2006). Figure 6.7 is a pictorial presentation of the four Cs of mission critical voice.

While there is a clear expectation that all first responders using MCPTT will be able to interoperate, this is less clear when discussing the use of different PTT solutions. A police officer using POC solution "A" must be able to communicate with a sheriff's deputy using POC solution "B." Direct mode PTT communications in LTE, which are conducted without any cellular infrastructure support, are also essential for public safety agencies and are a key safety component to protect first responders. However, LTE direct mode communications are still under development by 3GPP. Work is continuing on an LTE direct mode solution for public safety called "ProSe" (named for proximity based services). This is designed to support direct user to user voice, data, and video communications. In the meantime, industry is introducing alternative solutions in some public safety handsets including dual-mode user devices that support both LMR and on-network LTE communications and the provision of a narrowband voice radio within the LTE handset to provide direct mode capabilities. The introduction of various interim direct mode solutions may negatively impact long term interoperability. Public safety

Figure 6.7 The 4 Cs of mission-critical voice.

agencies will need a common LTE direct mode solution to achieve interoperable communications (Favraud and Nikaein 2015; Ferrús and Sallent 2015a,b,c).

The need for effective interoperability between LMR and LTE networks is now of paramount importance. Public safety agencies are likely to embrace the availability of existing POC solutions. These POC applications will receive priority in the NPSBN making them more reliable. In 2018, first responders will also have pre-emption capabilities that will further increase the reliability of these POC solutions. There are a variety of operational considerations and migration options that should be considered when contemplating the design of LMR-LTE interoperability.

- Interoperability may be needed within an agency to support the use of both LMR and LTE services that are assigned to different employees based on their agency role.
- Interoperability may be needed within an agency as it transitions all of their users from LMR to LTE.
- Interoperability may be needed by an agency to support interoperable communications with another agency on a disparate system (LMR or LTE).

According to a report by the National Public Safety Telecommunications Council (2018), the conclusion was that LMR and LTE interoperability will be required over an extended period of time. It is likely that some public safety agencies may never migrate their LMR operations over to LTE. First responders will also continue to need interoperability with other government and private entities that will remain on LMR networks. Public safety agencies should reference the SAFECOM Interoperability Continuum to assess all of the factors needed to achieve successful interoperability with LTE. While the Continuum was designed around LMR interoperability, the five "travel" lanes are also applicable to LMR-LTE interoperability. It is important to keep in mind that the report does not prescribe an LMR-LTE interworking solution. The Alliance for Telecommunications Industry Solutions (ATIS) and the Telecommunications Industry Association (TIA) has created a Joint LMR/LTE (JLMRLTE) subcommittee to study requirements for interworking of LTE mission-critical broadband services with existing P25 LMR systems.

While there are existing LMR-LTE interfaces in use today by public safety agencies, it is important that the standardized solution includes all of the requirements necessary for effective interoperability. LMR-LTE interoperability must include more than the exchange of voice traffic (Doumi et al. 2013). This report articulates the need for the exchange of various data elements, including PTT ID, Emergency Alert, and location information. Digital encryption is used by many public safety agencies today on their LMR networks and first responders will need a solution that supports end-to-end encryption during joint operations, where agency personnel are using both LMR and LTE interconnected talk groups. As noted earlier in this chapter, some solutions designed for LMR system interoperability may be leveraged to provide LMR-LTE interoperability. Finally, it should be recognized that public safety agencies come in all sizes and configurations. Many agencies face budgetary constraints and will not be able to afford feature-rich interoperability solutions. This requires the availability of a range of interworking solutions that meet the requirements of small agencies with basic needs as well as large agencies using complex systems, and one solution to this is the full establishment of 5G networks.

6.15 Convergence Solution for LTE and TETRA for Angola's National Communications Network

Angola took up the technological advancement step in 2017, which was provided by Hytera Company. The main client for this conversion project was the Ministry of Interior in Angola.

The project is a turnkey project, which has TETRA infrastructure, a LTE-TETRA, which is multimodal radios and TETRA radios, which have convergence systems.

6.15.1 The Objectives of the Project

i) This project is an initiative by the government of Angola to modernize public safety and security
ii) Modernization of the police service
iii) Modernization of the firefighting service
iv) Modernization of the Angolan Customs Department and other critical governmental departments.

This project utilizes a modern and new mission-critical system of communication. The TETRA-LTE convergence solution helps to ensure that the Angolan government has a highly resilient infrastructure that can sustain mission-critical users in the country, especially those providing critical services.

The convergence offers several versatile features that are collectively enabled by LTE technologies.

6.15.2 Advantages of the LTE-TETRA Solutions

Integration of LTE and TETRA as a unified public safety infrastructure is essential for public safety and mission critical users. It brings many advantages at different layers of applications, switching, and radios. The following are some of the advantages of this internetworking.

i) They are quick and effective at video streaming from the field because of their combined narrowband critical mission communication capabilities, which are possible through the LTE broadband data throughput capabilities
ii) Enable users to operate and acquire highly effective information because they support instant feedback, which in turn supports quick decision making
iii) They provide seamless communication among users
iv) They offer highly secure systems of communication between users because of the encryption in the equipment and channels of communication
v) Enhancement of data connectivity
vi) Creation of unification of the existing public safety networks that were once disparate
vii) It has enabled bringing together of all the disaster response teams using different frequencies together as it integrates technologies like TETRA, VHF, and UHF.

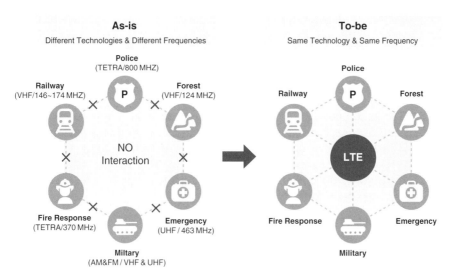

Figure 6.8 Network integration. Source: https://www.samsung.com/global/business/networks/solutions/public-safety-lte.

6.15.3 Illustration: Before Integration and After Integration

The fourth generation of mobile broadband has gained wide traction in the mobile networks industry and telecom players have developed products around its usage. For some time, LTE has become the fastest developing system available in the history of mobile communication. Wide application across different countries has ensured the global footprint necessary to usher lower costs owing to economies of scale. The telecom industry across different nations has vastly invested in LTE networks, achieving a unique combination in terms of quality of service and high data rates. From the perspective of public safety, LTE provides faster response times and high spectrum efficiency, which are critical in gaining interoperability across different networks. Therefore, the technical requirements espoused for the development of public safety networks are available in LTE if the technology for mission-critical providers can be built around the technology. Higher data rates are additionally available in the next step of LTE technology otherwise regarded as LTE-Advanced (Prasad et al. 2016, p. 42). Investing in the latter system would achieve higher functionality for public safety networks especially when it comes to sending data in times of emergency. Figure 6.8 shows LTE as a future unique platform for interoperability purpose.

6.15.4 Overview of LTE Technology

LTE technology has been the mobile technology standard developed since the 3GPP Release 8 with the latest being the 3GPP Release 13 otherwise referred to as LTE-Advanced Pro and primed to offer more functionalities and features. LTE technology has provided a flat architecture that is all internet protocol based (IP) and flexible in terms of spectrum utilization (Wang et al. 2014, p. 22). Additionally, compared to the radio communication standards used previously such as multiple in, multiple out (MIMO), 64QAM and 256QAM, turbo coding and carrier aggregation, the new standard attains superior spectrum efficiency with theoretical data rates of 150 Mbps (Khasdev 2016, pp. 19227–19229.). The system is further designed to achieve more spectrum efficiency beyond 400 Mbps. The development of LTE for commercial operators was only initially supposed to complement available GSM and UMTS networks but the desire for more efficiency is driving telecom operators to introduce voice, video, SMS, and internet over LTE technology, freeing up spectrum resources. With LTE having features that support broadcasting, the wide range of supported services would make a case for integrating public safety networks into the technology and develop around it for enhanced features (Doumi et al. 2013, p. 109). Although LTE does not offer the full portfolio of services required for use by public safety networks, developments are in process for:

- Group communication enablers according to 3GPP Releases 12 and 13
- Mission critical group call setup manageable in 300 ms with a maximum delay of 150 ms according to 3GPP Release 13
- Proximity services that are equivalent to those offered by TETRA direct mode operation (DMO) according to 3GPP Releases 12 and 13
- Isolated operations in disconnected areas from the main network for purposes of public safety according to 3GPP Release 13
- Mission critical transmission of video and data according to 3GPP Release 14.

PPDR networks of TETRA, TETRAPOL, and P25 would require being interoperable with the LTE network to get the full benefits of public safety networks evolution. Such interoperability should be on the interface level as well as in the PPDR narrowband networks. Achieving such interoperability would mean that public safety networks devices would connect to both PPDR networks as well as the commercial broadband networks (Ferrús et al., 2013).

6.16 5G Wireless Network and Public Safety Perspective

The fifth generation technology (5G) aims at transforming the wireless network infrastructure with a clustered network of small cells that employ speeds currently unattainable with 4G LTE technology. The economic growth and innovation have followed the infrastructural upgrades in the broadband ecosystem. Deploying a more connected system of communication devices has improved the speeds and has become the basis for introducing 5G over the available 4G LTE network. Compared to the number of available antennae and base locations in 3G and 4G LTE networks, the rollout of 5G will require ten to a hundred times more base stations for easier network resolution (Bader et al. 2016, p. e3101). A projected increase in the number of base stations will support the increased number of devices expected to be communicating to the networks bearing in mind the increasing adoption of mobile technology as well as a possible increase in virtual reality and the Internet of Things technologies.

The first development of broadband technologies associated applications has brought about emerging possibilities aimed at focusing on big data analysis, three-dimensional media, artificial intelligence as well as the Internet of Things requiring a significant amount of data. Mobile technology is already becoming indispensable to human beings with wide adoption of interactive applications. There is an increased focus on accessing broadband connections that are available anywhere and anytime making the progression into fifth generation (5G) and beyond (B5G) a matter of necessity. 5G mobile networks are designed to require enormous throughput per device in the ranges of Gbps with monthly global consumption predicted to hit 50 pb by 2021. Video transmission is the most dominant aspect of usage predicted and with the added focus of on-location videos; it is predicted to rise even higher. Mobile devices are increasingly requiring high definition videos necessitating the infrastructure to support such demands. The technical objectives required for 5G infrastructure would therefore be:

- Enormous high data rates computed per device
- Reliable support for a critical application like industrial controls, health care, and vehicle to vehicle communications
- High data rates per area and ensuring minimal interference across transmitters
- Low latency for virtual reality applications and interactive 3D videos

Integrating the 5G technology with existing TETRA, TETRAPOL, and P25 systems requires a degree of optimization to enable communication to straddle the two diverse technologies. This means that if the desired broadband technology is 5G, it would have the capability of co-existing with the available PMR systems. Presently, the evolution to integrate LTE-Advanced systems has yet to be optimized for voice calls and there is a likelihood for adopting the same approach in 5G to take advantage of the quality voice capabilities in PMR systems while optimizing the set up for enormous data solutions given by 5G.

6.16.1 Waiting for 5G for Public Safety Implementation

Mobile technological advances have become increasingly dynamic in the recent past with the 3G, 4G and 5G technologies having been introduced in the twenty-first century. Conventional wisdom would desire that public safety networks wait for resolution on the best technology to adapt to escape redundancy. Commercial broadband networks, however, have evolved with the evolution of technology without having to wait for such resolution. Such dynamism has enabled the proliferation of applications intended to take advantage of new technologies. Public safety networks require following the same script even as the short-termism may hurt the bottom line since there are enormous advantages of technologically moving with the seasons. The idea of waiting for the release of the next technological 3GPP standards may trigger a never-ending loop since the next technology is always an improvement of available technology. Committing to integrating PMR technologies with either 4G LTE networks or 5G technology would bring on board public safety networks to broadband technology, increasing the functionality and upgradability to emerging technologies. 5G is based on the concept of 4G LTE technologies ensuring that there is a progression of technology when PMR systems are brought on board. Changing the ecosystem from the narrowband PMR systems into the 3GPP would be a leap in the right direction irrespective of whether the targeted entry point would be 4G LTE or 5G technologies since it would allow the public safety networks to operate within the parameters of the upgradable broadband system (Bulusu et al., 2017).

6.17   The Linkage Between 4G and 5G Evolution

The deployment of 4G and 5G based on European and international regulators of the Electronic Communication Committee (ECC), International Telecommunication Union (ITU), and EC has mandated that the public safety spectrum should be in the 700 MHz band. Each country is, however, free to determine the exact range of applicable band in its territory. Since the rollout for 5G technology is still in its infancy, public safety networks would need to be deployed in the globally available 4G technology with the aim of bringing on board additional services as outlined in 3GPP Release 15 (Merwaday and Güvenç 2016, p. 984). The finalization of 3GPP Release 15 is scheduled for years 2018 and 2019 with the products utilizing the 5G technology expected 18 months later. Upgrading from 4G to 5G will only require a software upgrade within the 700 MHz band and utilizing the available infrastructure.

Deployment of 4G technology in some countries is currently ongoing even as the standards for 5G have been rolled out necessitating the application of part of 5G technologies in 4G networks in advance. If such were undertaken, the commercial users would benefit from the network evolution toward 5G technology while preparing to significantly improve applications necessary to improve spectrum efficiency (Chincholkar and Abdulmunem 2017, p. 552). With the expected ease of upgrading from 4G to 5G being a matter of software upgrades, the benefits of the 4G LTE network would be increased in 5G networks without the need for a redesign. Currently, the existing investments are protected in the changeover with ZTE 4G series of products, for example, supporting the smooth evolution from 4G to 5G while remaining relevant for 4G only networks. The convergence of the networks is consistent with the network evolution with the better approach being a systematic rollout.

The first stage toward a progressive evolution to 5G would require deploying 4G cutting edge products within the 4G networks together with 5G supporting infrastructure to achieve lower latency, more connections, and larger capacity. Progression from the first stage of the rollout to the second stage would require the deployment of innovative solutions that would support commercialization of the pre-5G solutions on a wider scale across countries. Convergence of 4G and 5G networks would be dependent on the dual deployment of systems that can harness the strengths of either systems or the infrastructure to support it.

6.17.1   Connecting 4G and 5G Solutions for Public Safety

The mobile broadband solution builds on the advantages of the preceding standards and is enhanced with more features. Progressively, the technology has evolved from 2G, 3G, 4G and now 5G with every improvement adding some additional features. 5G solution network implementations refer to 3GPP Release 15 onwards that include the deployment of 4G defined content as well as 5G specific content. No specific connectivity solution, therefore, would be required for exclusive 5G rollout as the 4G networks would only need a software upgrade to support the newer technology. 4G and 5G are only labeled defining the technological evolution from one level to another while retaining the capacity for interlinking between the technologies. Public safety networks that are therefore based on the 4G LTE networks would have the capability of accessing the 5G technology when the right software is put in place.

6.17.2 Deploying LTE Public Safety Networks

PPDR operators have varying choices when deciding on the most optimal deployment model that would serve the mission of their organizations and goals. Operations of critical mission organizations differ from one to another requiring an element of dynamism in the manner of upgrading to newer systems. Requirements needed by first responders, for example, significantly differ from those required by law enforcement officers in any area of jurisdiction. Deploying solutions for the different service providers should, therefore, consider the unique challenges to each organization. Designing the approach needed for deploying public safety networks within the fold of LTE technology would require broad-based solutions that encompass the unique challenges for each model and stratifying them across different options.

A dedicated PPDR would be one of the theoretical models that are based on the conventional approach regarding fully independent networks with equipment and infrastructure for each PPDR operator. Frequency spectrums, base stations, call networks, and operations would fully be owned by PPDR operators even as they ride on the new LTE network. Critical service organizations like the fire brigade, the ambulance and law enforcement officers across different states would need control rooms for their operations that are built according to the stringent requirements to ensure safety and reliability even as they ride on the LTE networks to share the envisaged benefits of newer platforms. Previously, the element of ownership was not regarded as being a challenge since narrowband PPDR networks operated on a relatively small prism. Bringing the networks to the LTE platforms would, however, require additional infrastructure that would take care of cost and complexity expected if each unit is to own the dedicated technology (Ferrús and Sallent 2015b).

A shared option for PPDR networks but riding on the LTE technology is a viable option for integrating public safety networks into the broadband technology. Such would be shared with commercial mobile operators with specifications on which aspects of the network would be shared, allowing the public safety networks to ride on the newer technology without compromising accessibility and reliability (Wei et al. 2011). Progressively, the public safety networks can agree to share part of the LTE based networks with the commercial mobile operators while using some of the PPDR networks on voice and short messaging services while working out on the right modality for fully embracing LTE networks without compromising safety and reliability.

6.18 Conclusion

Public safety networks perform a critical function of ensuring safety to citizens across different countries. The manner of communication and deployment of different personnel is dependent on technologies that have not evolved with the evolution of broadband technologies. Desiring to leverage the advantages brought about by LTE based solutions, the public safety networks have to find a solution that works to enhance their operations within the parameters of newer technologies. Such ventures as stipulated in this chapter have to be guided by the desire to firewall the systems against intrusion by the public in the same way they have operated with conventional technologies. Newer technologies in the context of the public safety domain are available for such utilization but ensuring that they are tweaked to ensure priority, safety, and reliability (Abrahão and Vieira 2016,

p. 427). Sharing of the spectrum should not mean a compromise in the way the public safety networks operate and sufficient safeguards should be put in place to ensure the technologies benefit all concerned parties.

References

Adibi, S., Agnew, G.B., and Tofigh, T.B. (2008). End-to-end (e2e) security approach in wimax: a security technical overview for corporate multimedia applications. In: *Handbook of Research on Wireless Security* (eds. Y. Zhang, J. Zheng and M. Ma). Information Science Reference.

Ahmad, I., Chen, W., and Chang, K. (2017). LTE-railway user priority-based cooperative resource allocation schemes for coexisting public safety and railway networks. *IEEE Access* 5: 7985–8000. Web.

Bader, F., Martinod, L., Baldini, G. et al. (2016). Future evolution of public safety communications in the 5G Era. *Transactions on Emerging Telecommunications Technologies* 28 (3): e3101. Web.

Balachandran, K., Budka, K.C., and Chu, T.P. (2006). Mobile responder communication networks for public safety. *IEEE Communications Magazine* 44 (1): 56–64.

Bartoli, G., Fantacci, R., Gei, F. et al. (2017). Dynamic PSS and commercial networks spectrum sharing in the time domain. *Transactions on Emerging Telecommunications Technologies* 28 (3): e3083.

Baskaran, S., Raja, G., Bashir, A. et al. (2017). Qos-aware frequency-based 4G+relative authentication model for next generation LTE and its dependent public safety networks. *IEEE Access* 5: 21977–21991. Web.

Bulusu, S.C., Renfors, M., Yli-Kaakinen, J. et al. (2017). Enhanced multicarrier techniques for narrowband and broadband PMR coexistence. *Transactions on Emerging Telecommunications Technologies* 28 (3): e3056.

Chavez, K.M.G., Goratti, L., Rasheed, T. et al. (2015). *The Evolutionary Role of Communication Technologies in Public Safety Networks*, 21–48. Elsevier.

Chincholkar, Y.D. and Abdulmunem, M.L. (2017). Hybrid approach of spectrum-energy efficiency optimization for LTE network. *IJARCCE* 6 (6): 549–554. Web.

Doumi, T., Dolan, M.F., Tatesh, S. et al. (2013). LTE for public safety networks. *IEEE Communications Magazine* 51 (2): 106–112. Web.

Dunlop, J., Girma, D., and Irvine, J. (2013). TETRA system architecture, components and services. In: *Digital Mobile Communications and the Tetra System*, 161–204. Wiley.

El Alaoui, S. (2012). Towards future 4G mobile networks: a real-world IMS testbed. *International Journal of Next-Generation Networks* 4 (3): 31–43. Web.

El-Keyi, A. et al. (2017). LTE for public safety networks: synchronization in the presence of jamming. *IEEE Access* 5: 20800–20813. Web.

Favraud, R. and Nikaein, N. (2015). Wireless mesh backhauling for LTE/LTE-A networks. In: *Military Communications Conference, MILCOM 2015–2015 IEEE*, 695–700. IEEE.

Fereidountabar, A. (2013). An improved DCA and channel modelling algorithm based on carrier priority in LTE/LTE-A. *American Journal of Networks and Communications* 2 (6): 149. Web.

Ferrús, R. and Sallent, O. (2015a). *Mobile Broadband Communications for Public Safety: The Road Ahead through LTE Technology*. Wiley.

Ferrús, R. and Sallent, O. (2015b). Future mobile broadband PPDR communications systems. In: *Mobile Broadband Communications for Public Safety: The Road Ahead through LTE Technology*, 81–124.

Ferrús, R. and Sallent, O. (2015c). LTE networks for PPDR communications. In: *Mobile Broadband Communications for Public Safety: The Road Ahead through LTE Technology*, 193–256.

Ferrus, R., Pisz, R., Sallent, O. et al. (2013). Public safety mobile broadband: a techno-economic perspective. *IEEE Vehicular Technology Magazine* 8 (2): 28–36.

First Responder Network Authority (2012). Why Firstnet, [Available online at http://www .firstnet.gov/about/why]

Frosali, F., Gei, F., Marabissi, D. et al. (2015). Interoperability for public safety networks. In: *Wireless Public Safety Networks 1*, 127–162.

Guo, K. and Rigelsford, J.M. (2013). A broadband tunable band-pass filter for RF characterization within TETRA and LTE bands. *Microwave and Optical Technology Letters* 55 (12): 2847–2849. Web.

Guoru, D., Yutao, J., Jinlong, W. et al. (2018). Spectrum inference in cognitive radio networks: algorithms and applications. *IEEE Communications Surveys & Tutorials* 20 (1): 150–182.

Kable (2016). First responder solutions in the UK and internationally. *International Journal Of Engineering And Computer Science* 5 (11): 19227–19229.

Khasdev, B. (2016). Overview of MIMO technology in LTE, LTE-A & LTE-A-pro. *International Journal of Engineering and Computer Science*. Web.

Kumbhar, A., Koohifar, F., Guvenc, I., and Mueller, B. (2017). A survey on legacy and emerging technologies for public safety communications. *IEEE Communications Surveys & Tutorials* 19 (1): 97–124.

Mason, A. (2010). Report for the TETRA Association Public safety mobile broadband and spectrum needs Final report. TCCA

Merwaday, A. and Güvenç, I. (2016). Optimisation of Feicic for energy efficiency and spectrum efficiency in LTE-advanced Hetnets. *Electronics Letters* 52 (11): 982–984. Web.

Munir, D. et al. (2015). Reliable cooperative scheme for public safety services in LTE-A networks. *Transactions on Emerging Telecommunications Technologies* 28 (2): e3008. Web.

National Public Safety Telecommunications Council 2018 Public safety Land Mobile Radio (LMR) Interoperability with LTE Mission Critical Push to Talk, NPSTC, Final Report National Public Safety Telecommunications Council http://www.npstc.org/download .jsp?tableId=37&column=217&id=4031&file=NPSTC_Public_Safety_LMR_LTE_IO_ Report_20180108.pdf

Naqvi, S.A.R. et al. (2018). Drone-aided communication as a key enabler for 5G and resilient public safety networks. *IEEE Communications Magazine* 56 (1): 36–42. Web.

Oueis, J., Conan, V., Lavaux, D. et al. (2017). Overview of LTE isolated E-UTRAN operation for public safety. *IEEE Communications Standards Magazine* 1 (2): 98–105.

Paulson, A. and Schwengler, T. (2013, September). A review of public safety communications, from LMR to voice over LTE (VoLT E). In: *2013 IEEE 24th International Symposium onPersonal Indoor and Mobile Radio Communications (PIMRC)*, 3513–3517. IEEE.

Prasad, A. et al. (2016). Enabling group communication for public safety in LTE-advanced networks. *Journal of Network and Computer Applications* 62: 41–52. Web.

Ragaleux, A., Baey, S., and Gueguen, C. (2015). Adaptive and generic scheduling scheme for LTE/LTE-A mobile networks. *Wireless Networks* 22 (8): 2753–2771. Web.

Stavroulakis, P. (2007). *Terrestrial Trunked Radio-TETRA: A Global Security Tool.* Springer Science & Business Media.

Stojkovic, M. (2016). *Public safety networks towards mission critical mobile broadband networks* (Master's thesis, NTNU).

Tata, C. and Kadoch, M. (2014). Efficient priority access to the shared commercial radio with offloading for public safety in LTE heterogeneous networks. *Journal of Computer Networks and Communications* 2014: 1–15. Web.

TCCA (2018), TETRA Release 2. Newcastle upon Tyne: TCCA Available online at http://www.tandcca.com/about/page/12029. [Accessed 20 June 2018]

Usman, M. et al. (2015). A software-defined device-to-device communication architecture for public safety applications in 5G networks. *IEEE Access* 3: 1649–1654. Web.

Vilhar, A., Hrovat, A., Ozimek, I., and Javornik, T. (2017). Analysis of strategies for progressive 5G emergency network deployment. *Transactions on Emerging Telecommunications Technologies* 28 (3): e3059.

Wang, J. et al. (2014). Issues toward networks architecture security for LTE and LTE-A networks. *International Journal of Security and Its Applications* 8 (4): 17–24. Web.

Wei, Q., He, J., and Zhang, X. (2011). Privacy enhanced key agreement for public safety communication in wireless mesh networks. *Journal of Networks* 6 (9). Web.

7

Bringing Public Safety Communications into the 21st Century

7.1 Emerging Technologies with Life-Saving Potential

In the present day, life-saving technology has steadily gained momentum worldwide and remains revolutionary due to the fact that it empowers even a layman to become easily knowledgeable and a hero in one way or another. Nevertheless, a number of new life-saving technologies are still undergoing development that have the capacity to provide immense promise for first incident responders and bystanders (Zarrilli 2016). Emerging technologies have become part of life, simply because some people may interpret these technologies as emerging, and yet others may see them in different ways (Conway 2013). BusinessDictionary.com defines emerging technology as a new technology that is still under development or being developed to be implemented at a future time, in about five to ten years (BusinessDictionary 2018). This new technology will considerably transform or change the social and business environment, including communication capabilities, and will alter the traditional way of conveying information. The technology will entail wireless data communications, information technology, biotechnology, on-demand printing, advanced robotics, and man–machine communications (Conway 2013).

Nancy Torres, an expert and writer for a government technology magazine revealed that The National Oceanic and Atmospheric Administration indicated that the year 2017 should be remembered as one of the most expensive times in United States records for disaster, showing that approximately $306 billion was destroyed in damage. Similarly, the US Federal Bureau of Investigation (FBI) also reported the same year as having the highest number of cases of shootings, with the majority of people killed in just one year by gun men or active shooters (Torres 2018). As a result, with such an increase in crises and potential damage across the US, technology and data play a central role in enhancing and supporting emergency management departments all around the country and the world at large. It is reported that about 240 million different calls are made to 911 departments in the US each year, with at least 80% of them originating from wireless technology, and yet most of the emergency management coordination continue to operate on legacy networks designed and installed for wireless phones. Consequently, the most affected population in need cannot share precise physical locations or addresses or even send critical messages via media platforms to responders, which in turn makes communications within the emergency communication department and resource allocations more difficult and costly to the government (Torres 2018).

Public Safety Networks from LTE to 5G, First Edition. Abdulrahman Yarali.
© 2020 John Wiley & Sons Ltd. Published 2020 by John Wiley & Sons Ltd.

Federal governments and city entities are significantly making possible steps in using analytics to enhance emergency response operational activity, from using Google's 911 analyses in San-Francisco to the Department of Management and Budget's operational work with the Federal Emergency Management Agency (FEMA) to have more sources flowing in real-time information, especially during critical emergencies. Nevertheless, emergency technologies bring considerable opportunities to improve and support the emergency management network to make it more secure, effective, and intelligent. This comes in a time when most federal governments and some cities have for some time wanted to integrate high-tech into disaster preparedness and response, and so current technologies and capabilities are increasingly evolving every day. Most importantly, today the Internet of Things (IoT), blockchain technology, and artificial intelligence (AI) all offer the potential to improve public safety as they play a big role in generating, transmitting, and reading emergency-related data to have powerful decision-making in future crises (Torres 2018).

7.1.1 Artificial Intelligence

Artificial Intelligence, which is normally abbreviated as AI, is basically a term used in the field of technological machines attributed to simulated intelligence. Artificial intelligence in a machine functions in a way that a machine is programmed to mimic the way a person acts and also to think like a human (Dickens and Cook 2006). AI is characterized with the ability to rationalize and take actions that have the best chance of an achieving a certain particular goal. Artificial intelligence, significantly, is mainly based on the idea that the intelligence of a human can be defined in the precise terms that the machine is capable of mimicking. Artificial intelligence has basic goals that include reasoning, learning, and perception. The following are explanations about how AI has lifesaving potential in an advancing technology; although there has emerged a number of controversies about its usage and applicability and effects of artificial intelligence in people's future lives, it has certain potential for life saving. The ways in which AI has lifesaving potentials are discussed as follows.

That first all artificial intelligence is a particular software built to learn or problem solve processes that are performed in the brain of a human. The other wonderful thing about AI is that it can even create visual art or write songs. Artificial intelligence also has the potential to be revolutionary; it has the capability of changing the way people work through replacing humans with machines and software, leading to developments in the areas of self-driving cars hence making driving a thing of the past. That way the high risk of accidents occurring on the roads caused by cases of fatigue and loss of control, as a result, claiming people's lives, will be brought to an end or lessened. People's methods of shopping could also be changed by artificial intelligence shopping assistants, thereby relieving pressure on people who have such busy schedules that they hardly get time to do important things like shopping.

Advancement in medicine through the introduction of robotic surgery is another way in which AI has helped save lives. Robotic surgery is defined as "a method to perform surgery using very small tools attached to a robotic arm." In 1996, David Burrett who was then a doctoral student at Massachusetts Institute of Technology built a biometric robot called robot Tuna with the aim of using it to study how fish swim in water. It was designed in the form of fish. In 2000, Honda introduced a humanoid project called ASIMO. It was a type of robot that was capable of communicating, walking, running, and interacting with humans; it also had the ability of environmental and facial recognition. Another type of robot was launched in April 2001 known as Canadarm2, which was attached to the international space station. This robot is a larger version of an arm that is used in the space shuttle and in 2004. It is a robot that has self-replication capabilities and was invented by Cornell University.

In a remote medical procedure, for example, the surgeon controls the robotic arm with a computer (US National Library of Medicine 2013). General anesthesia is applied to the patient to ensure that the procedure is painless; the surgeon sits in front of a computer station to view the entire surgical procedure. Small surgical tools are attached to the robot's arms; small incisions on the patient's body parts on the surgical site are made as the entry point of the instruments. A thin tube with a camera attached to the end of it (endoscope) allows the surgeon to view enlarged 3D images of your body as the surgery is taking place (U.S. National Library of Medicine 2013, p. 1); and the robot allegedly matches or replicates the surgeon's hands during the operation. The emergence of robotic surgery stems from revolutions under the biomedical field. The procedure falls within the minimally invasive surgery approach (Brown University 2005). Litigation is complex due to the liabilities that need to be evaluated in terms of participation of three parties: the physician, the hospital, and the maker of the robotic system. Other factors to evaluate include: equipment safety and reliability, provision of adequate information, and maintenance of confidentiality are all of paramount importance. It identifies that apart from being quick at carrying out surgeries, the robots operated in a very precise and concise manner with limited levels of errors and inaccuracies as they are programmed to follow predetermined steps. It further explains the possible risks that may be associated with this kind of surgery similar to those that may be encountered in conventional surgical operations. The major setback to this kind of surgery is that the equipment is not readily available to all the patients and it is extremely expensive to maintain this equipment. Finally, the cost of robotic surgery and the lack of such systems in most of the public hospitals may restrict the majority from the benefits offered by the new technology (Mavroforou et al. 2010). Dickens and Cook (2006) identify the legal aspects surrounding the use and practice of this kind of surgery. They explain the need for the patient's consent before the method of robotic surgery is applied to patients. They further explain the legal penalties to a practitioner who wrongfully or unlawfully applies this kind of surgery without the patient's permission. It provides the legal framework and backing related to this kind of surgery. It explains that practitioners that employ this kind of surgery ought to be registered and licensed to conduct it as it violates the law for them to apply it if not licensed.

Lanfranco et al. (2004) put emphasis on the movement and developments that have been associated with the concept of robotic surgery. The resource further presents the current perspective, especially the extent to which this kind of surgery is being applied today in relation to current advances in the technologies that are associated with this kind of technology. All the historical developments in this kind of surgery right from the original authors and developers of this method surgery are identified in the above resource. In addition, they identified some of the advances that have been made in this kind of surgery to ensure that it is a success and how it is set to replace the conventional methods of surgery.

The doctors that endeavor to carry out this kind of operation are faced with a very steep learning curve. However, using the robots in such operations, precision, and accuracy are attained without the need to have the robots learn as they just have to be programmed on set repetitive commands and the procedures are done. The resources also explain that those surgeries can be carried out virtually employing the method of telescoping. Using this method an operation can be carried out by someone or a robot or robots taking instructions from someone away from the patient. There is the tremendous success that is associated with surgeries that are related to the human alimentary canal, especially the colon and rectum, using minimally invasive methods employing robots. Such operations do not require the individual's alimentary canal or body to be openly dissected but rather using a small intrusion that can facilitate robotic surgery. In addition, the laparoscopic and robotic operations have enjoyed tremendous success in performing these complex operations in patients.

As a result of artificial intelligence, a prosthetic head was invented. The prosthetic head refers to an artificially created head known as the embodied conversational agent that is made and programmed in such a way that it is able to speak and respond to the person who may be interrogating it. With this kind of advancement in the conversational technology industry, it has been made easy for people to conduct transactions, obtain information and as well be entertained by the simple means of just speaking to a computer that is programmed to give the necessary information. On the basis of dialogue understanding and integration of speech recognition among others, conversation systems are fast becoming important and significant components of customer care, solutions, and virtual assistance, thereby saving money and time in terms of easy access to information as well as attracting customers. This, therefore, leads to positive effects on the development of the industry and promoting innovations (Tear et al. 2016).

Finally, the CEO of Woolah, Matt Shames, once stated that beginning from 2018 onwards, the world would see rapid consolidation of AI and machine learning technology. That now with machine learning, a person can accurately classify what is present in a photo, including all its engagement data such as comments, and emotions as well as the activity it represents in the image. This effectively optimizes human-generated content to reach its true potential. It has also been said that AI, as technology advances, will replace repetitive tasks. This is according to the founder of Artivatic Data Labs (Artivatic, ai), Layak Singh. He stated that AI is utilized in various sectors such as education, medical, logistic, etc. The technology will enable systems, operations, and processes to have thinking power like the human brain to make self-decisions without human intervention. He further stated that some of the critical areas, such as diagnostics, supply chain, education, and manufacturing, will touch new heights toward making human lives easier(Entrepreneur India 2018). Based on the information about artificial intelligence, one individual could then say that there is no doubt that advancements in technology have made people's lives better, convenient, and easier. That people in today's generation are lucky to have to leave in an era where there are new emerging technologies that also impact positively on people's lives.

7.1.2 The Internet of Things (IoT)

A single thing on the IoT can be an animal on a farm with a biochip transponder, a person with a heart monitor implant, an automobile that has built-in sensors to notify the owner or driver about any unseen engine operations or when the tire pressure is low, and any natural or man-made object that can be assigned a unique internet protocol (IP) address and can transfer required data over the network. Today, most business organizations have embraced this technology and are currently using IoT to operate more securely and effectively, and understand their customers to have improved clientele service, enhance decision-making processes and improve business value and benefits (Rouse 2018). According to Bornheim and Fletcher (2016), IoT is a new concept in business and technology fields where people and devices including appliances all become connected end-points in broader complete internet connectivity as well as a network of networks. In other words, it refers to the design and use of internet based networks and applications that directly interact with the environment (Bornheim and Fletcher 2016). In simple terms, the technology creates significant expectations that the internet network is always turned "on" and that all devices, now become "everything" is or can be connected either directly, using an adaptor, or through using an application on a device, such as apps installed on a smartphones, tablets, iPads, and proxying the device itself.

In other words, sometimes IoT is defined as machine to machine (M2M) or human to machine (H2M) method of communication. Although the definition varies depending on who is saying it or the business setting, M2M connection is estimated to significantly grow from 4 billion from 2014 to 26 billion in 2025 and beyond that (Bornheim and Fletcher 2016; Rouse 2018). This trend implies that telecom and information and communication technologies (ICTs) firms see and project enormous revenues, and as a result, the company can expect a huge volume of developments to be generated from that line, and more new business approaches will be introduced more often (Bornheim and Fletcher 2016). In short, the IoT refers to a complete system of interrelated computing devices, digital and mechanical devices, animals, natural or man-made objects, and/or people that are offered with an IP address or unique identifier, which has the capability to transfer data over the internet or network of networks, without employing H2M or M2M interaction (Rouse 2018). On the other hand, the IoT is defined as the network of tangible objects embedded with software and sensors that collect required data and interact with each other. The IoT technology relates to the emergency management department: it can be implemented by industry organizations to improve data collection from different physical settings and easily convey this data to various departments (Torres 2018).

In most cases, weather-related disasters including floods and hurricanes tend to prevent emergency response departments from reaching and entering some hazardous places. Such hindrances largely reduce the response team's capacity to track all damage, alert the concerned or responsible public with relevant and updated information regarding the incident, and respond in the shortest time possible. Nevertheless, if IoT machines were installed in such places, they would easily be able to send signals and communicate important data, including water quality, smoke, and temperature (Torres 2018). Having this data, governments can come up with a better informed decision on how to allocate available resources, especially during catastrophe situations. Currently, in the city of Rio, the Hall Operations Center uses computerized sensors to gather real-time data concerning police, traffic, weather forecast, and health services in the city. It is also reported that in the US, Houston city works together with AT&T in the aftermath of Hurricane Harvey to employ IoT for identifying injuries and communicating critical data (Torres 2018). With this trend, governments and cities are able to implement IoT on the public safety infrastructures to monitor the potential risk factors and communicate critical data concerning possible emergencies. For instance, The Lower Colorado River Authority (LCRA) has installed a total of 20 electronic sensors to monitor and measure how fast water is flowing over a stream and shows what the water is doing at all points. With this, LCRA has the capacity to proactively monitor and control floods and quickly get to the bottom of water-related catastrophes in the area (Torres 2018).

7.1.3 Blockchain

Blockchain is a new technology and being developed; it has become a tool that many claim has the capacity to transform how business organizations transact data. It is an indisputable and disseminated digital ledger, secured by cryptography that can be programmed to capture a number of business transactions. With numerous investments over $1 billion in only 2015, the social and economic impact of blockchain technology's potential to significantly transform the way governments and business organizations operate is now emerging (Cann 2016). Today, one of its most scalable apps is Bitcoin technology, a cryptocurrency and online payment network, which is gaining much attention across the world. The role of blockchain in emergency situation management is that has the ability to offer transparency and interoperability. In terms of interoperability, business organizations can implement it as a universal system across departments, similar to the internet, and enable a number of entities to interact across the system to allocate available resources in an emergency situation (Torres 2018).

For example, in the event of disaster relief, different parties join and often contribute possible resources to aid the affected places. Meanwhile, if these party entities engaged in these incidents were to embrace and implement a blockchain based shared network of data, they would quickly bring together potential efficient disaster responses to ensure that all resources were delivered to the most affected places and where they are needed the most. In the United States the Centers for Disease Control and Prevention (CDC) has embraced blockchain technology for the use-case of public health and safety data inspection, where the system gathers and conveys the required information to parties that engage in providing medical treatment to patients in disaster relief situations, such as pharmacies, hospitals, and public health agencies (Stanley 2017).

Within the disaster relief situations, blockchain has been tested to offer indisputable data that is available to everyone, to demonstrate what resources have been streamlined to be allocated to a certain area and by who. This transparent data could minimize the possibility of mismanagement of resources and corruption in some governments (Torres 2018).

UNICEF is also piloting blockchain technology to monitor the status of international grants in a more secure manner that is available to the public. Similarly, FEMA's Public Assistance project would be another potential use-case for this technology, tracking destinations in the aftermath of a disaster (Hochstein 2017). A number of factors tend to hinder and prevent the progress of blockchain from growing across any organization today, but emergency management departments across governments and cities should join this trend to explore more about this technology and its apps to better plan for potential information technology systems (Torres 2018).

References

Bornheim, M., & Fletcher, M. (2016). *Public safety digital transformation. The Internet of Things (IoT) and Emergency Services. EENA Technical Committee Document, Brussels.* [Available Online at http://www.eena.org/download.asp].

Brown University. (2005). Robotic Surgery. Retrieved from biomed.brown.edu: http://biomed.brown.edu/Courses/BI108/BI108_2005_Groups/04

BusinessDictionary. (2018). emerging technologies. Retrieved from businessdictionary.com: http://www.businessdictionary.com/definition/emerging-technologies.html

Cann, O. (2016, June 23). These are the top 10 emerging technologies of 2016. Retrieved from weforum.org: https://www.weforum.org/agenda/2016/06/top-10-emerging-technologies-2016

Conway, D. (2013, April 29). What Is Emerging Technology? Retrieved from Stevens-Henager.ed: https://www.stevenshenager.edu/blog/what-is-emerging-technology

Dickens, B.M. and Cook, R.J. (2006). Legal and ethical issues in telemedicine and robotics. *Int. J. Gyn. Obster.* 94 (1): 73–78.

Entrepreneur India. 2018 *How Artificial intelligence Technology will Change our Lives.* Retrieved on 30/11/2018 from: https://www.entrepreneur.com/article/306284

Hochstein, M. (2017). Perspectives On Disaster Recovery: How Blockchain Technology Can Make The Delivery Of Fema Public Assistance More Effective. Retrieved from hagertyconsulting.com: http://hagertyconsulting.com/about-us/blog/4527-2

Lanfranco, A.R., Castellanos, A.E., Desai, J.P., and Meyers, W.C. (2004). Robotic surgery: a current perspective. *Annals of Surgery* 239 (1): 14.

Mavroforou, A., E, M., Hatzitheo-Filou, C., and Giannoukas, A. (2010). Legal and ethical issues in robotic surgery. *International Journal of Angiology* 29 (1): 75–79.

National Library of Medicine. (2013). Robotic surgery. Retrieved from nlm.nih.gov: https://www.nlm.nih.gov/medlineplus/ency/article/007339.htm

Rouse, M. (2018, June). Internet of Things (IoT). Retrieved from internetofthingsagenda.techtarget.com: https://internetofthingsagenda.techtarget.com/definition/Internet-of-Things-IoT

Stanley, A. (2017). Centers for Disease Control to Launch First Blockchain Test on Disaster Relief. Retrieved from coindesk.com: https://www.coindesk.com/us-centers-disease-control-launch-first-blockchain-test-disaster-relief

Tear, M., Zoraida, C., and Griol, D. (2016). *The Conversational Interface: Talking to Smart Devices.* Springer Publisher.

Torres, N. (2018) Justice and Public Safety: Three Emerging Technologies with Life-Saving Potential. Retrieved from govtech.com: http://www.govtech.com/public-safety/Three-Emerging-Technologies-with-Life-Saving-Potential.htm

Zarrilli, Z. (2016, February 8). 7 Life-Saving Technologies You Need to Know About. Retrieved from surefirecpr.com: https://www.surefirecpr.com/7-life-saving-technologies-need-know

8

4G LTE: The Future of Mobile Wireless Telecommunication Systems for Public Safety Networks

8.1 Introduction

There have been many changes in the way telecommunication companies have been doing business – not only to satisfy the needs of the consumer but also to compete with other telecommunication services. Advances in technology have led to the integration of several telecommunication services and devices. Improved semiconductor and electronics manufacturing technology, and the growth of the internet and mobile telecommunications are some of the factors which have fueled this growth in telecommunications. One of the biggest concerns for wireless service providers is that of coping with a growing demand for data. Users are consuming more data every year and the telecoms industry over the last few years has seen unprecedented mobile data traffic growth due to the increasing adoption and usage of smartphones, tablets, and many other small handheld devices. As the reliance on the internet continues to grow, so too will the demand for higher data capacity and throughput. In addition, the number of mobile subscribers is also growing, making service providers worry about how to most efficiently allocate their available spectrum. Overloading service cells and having insufficient data capacity and throughput will continue to be a pressing concern. Long Term Evolution (LTE) was designed as an upgrade to the 2G and 3G systems that were in use previously. In an attempt to meet the standards proposed by the International Telecommunication Union (ITU), LTE includes several specifications and features that are far superior to pre-LTE technologies, while still falling short of the outlined ITU specifications. Some of the key technologies used in LTE include multiple in, multiple out (MIMO), coordinated multipoint transmission (CoMP), relay node (RN), and self-optimizing network (SON). LTE-Advanced, on the other hand, improved significantly upon LTE, to the point that the ITU has labeled it as "true 4G." The benefits of implementing LTE-Advanced are almost limitless. It has much higher throughput and bandwidth due to high-order MIMO, carrier aggregation (CA), and flexible spectrum usage. In this chapter, we discuss the incorporation of several new small technologies and multiple dimensions of benefits contributed to resolving staggering high throughput and capacity demands posed on wireless networks. We present some of the simulation results for these small technologies in LTE and discuss issues and concerns of service providers and customers. We also present how LTE can bring public safety (PS) networks to an interoperable and efficient infrastructure.

Public Safety Networks from LTE to 5G, First Edition. Abdulrahman Yarali.
© 2020 John Wiley & Sons Ltd. Published 2020 by John Wiley & Sons Ltd.

The challenges facing modern telecommunication system operators differ by region, by company and marketplace size, and by their area of focus and differentiation. However, there are certain themes such as revenue preservation, revenue expansion, brand extension, operational efficiency and speed-to-market cycles that impact all of them. Mobile operators are shifting their infrastructure planning toward software to become more agile. Carriers need to invest in intelligence, from core to backhaul to the network edge, not just in hardware capacity, in order to deliver quality services. It can be claimed that telecommunications science and its application in commerce and industry was the most rapidly evolving area of technology in the last decade. The internet grew exponentially during this period, in both the number of users and the amount of content available. Developed countries now boast high speed connections with a large percentage of homes having access to the internet and broadband services at an affordable fee. Underdeveloped countries, however, are yet to enjoy such facilities. This is referred to as the digital divide. The digital divide was defined as the unequal access to information and communication technologies (ICTs), where the least developed countries are separated from the developed countries because of a lack of technology, particularly ICT.

The cellular industry began developing 2G systems in the early 1980s. As experience shows, the lead time for mobile phone system development is about 10 years. Primary thinking on 3G took place in 1991 as 2G (global system for mobile communications, GSM) systems just started to roll out. Therefore, it was felt that 4G should have been operational from around 2011, and built on the second phase of 3G when all networks are expected to embrace internet protocol (IP) technology. In 2005, companies such as Ericsson, Motorola, Lucent, Nortel, and Qualcomm came up with "3G-plus" concepts that would push the performance of approved, though still emerging, standards beyond current ones. LTE and later on LTE-Advanced were designed as an upgrade to the 2G and 3G systems that were in used previously. In an attempt to meet the standards proposed by the ITU, LTE includes several specifications and features that are far superior to pre-LTE technologies, while still falling short of the outlined ITU specifications. Figure 8.1 shows a roadmap of the evolution of wireless technologies over the last two decades.

The ongoing development of mobile devices and their applications increases the requirements for high data rates and large capacity of the wireless communication networks rapidly. Wireless operators have already been overwhelmed by the exponential growth of data traffic on their networks and every service provider must ensure that its network can respond to the exponentially growing traffic, see Figure 8.2.

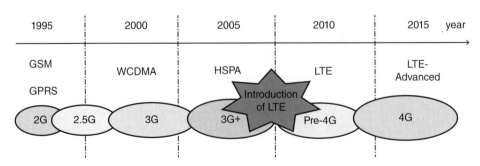

Figure 8.1 Migration path of wireless technologies.

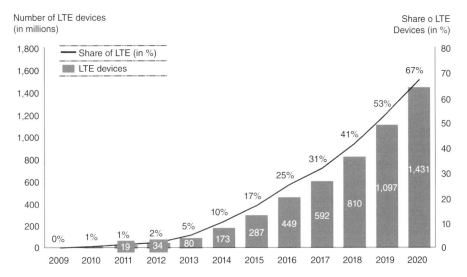

Figure 8.2 Number of mobile devices shipped worldwide annually and share of LTE devices. Source: Ovum mobile devices forecast, 2008; Booz & Company.

LTE offers much greater performance at much lower cost due to its spectral efficiency and it is expected to be able to accommodate that fast and staggering growing demand for wireless data services for the future. 4G LTE circumvents this issue first by having an all packet-switched network. Furthermore, it runs on the same networking system as most modern networks, including the internet – IP. This means even voice data, traditionally carried via circuit-switching, is now also transported via packet-switching. This is accomplished via VoLTE (voice over Long Term Evolution) (Lawson 2012). This actually allows for better sound quality and should also allow the 3G spectrum to be reused. The available bandwidth is also very wide, ranging from 1.25 to 100 MHz for all LTE technologies combined. This will serve to integrate several other services such as video conferencing and instant messaging services alongside the PS voiced data. This service class is known as the rich communications service (Smith 2010).

LTE, which was developed as part of the Third Generation Partnership Project (3GPP), is capable of providing high peak data rates (10 Mbps downlink and 50 Mbps uplink), multi-antenna support, reduced cost, wide range of bandwidth (from 1.4 up to 20 MHz) (from 1.24 to 100 MHz) (Lawson 2012), enabling a round trip of less than 10 ms, backward compatibility with existing 2G and 3G networks, increased spectrum efficiency and peak data rates at cell edges. All these criteria are satisfied by the efficient usage of the control channels. LTE is said to be an updated version of the Universal Mobile Telecommunications System (UMTS) technology, which gives it the ability to offer faster data rates for downloading and uploading and meets the requirement of 4G set by ITU. These requirements are as follows:

1. Be based on an all-IP packet-switched network.
2. Have peak data rates of up to approximately 100 Mbps for high mobility such as mobile access and up to approximately 1 Gbps for low mobility such as nomadic/local wireless access.

3. Be able to dynamically share and use network resources to support more simultaneous users per cell.
4. Reduced cost per bit.
5. A reduction in terminal complexity with an allowance for reasonable power consumption.
6. Using scalable channel bandwidths of 5–20 MHz, optionally up to 40 MHz.
7. Have peak link spectral efficiency of 15 bps/Hz in the downlink, and 6.75 bps/Hz in the uplink (meaning that 1 Gbps in the downlink should be possible over less than 67 MHz bandwidth).
8. System spectral efficiency of up to 3 bps/Hz/cell in the downlink and 2.25 bps/Hz/cell for indoor usage.
9. Smooth roaming and handovers across heterogeneous networks.
10. The ability to offer high quality of service (QoS) for next generation multimedia support.

Due to the need to use different bandwidths and data rates, LTE uses orthogonal frequency division multiple access (OFDMA) in the downlink and single carrier frequency division multiple access (SCFDMA) is frequently used in uplink. LTE differs from its predecessors by using orthogonal frequency digital multiplexing (OFDM) along with MIMO antennas. OFDM is selected because of its suitability for MIMO transmission and reception, resistance of its symbol structure to multi-path delay spread, no need for equalization, etc. LTE supports both time division duplexing (TDD) and FDD frequency division duplexing. FDD uses two separate frequency bands to allow transmission and receiving simultaneously from both sender and receiver. TDD on the other hand uses a single frequency band to send and receive simultaneously from both devices.

While LTE is still relatively new, there has been some enhancements made to the system. One of the main candidates to fulfill the IMT-Advanced requirements is the Long Term Evolution-Advanced (LTE-A) system. New technologies included in LTE-A systems include multi-antenna systems for MIMO uplink and downlink, coordinated multi-point transmission and reception (CoMP), bandwidth extension using CA, the use of repeaters to improve cell edge capacity and the deployment of heterogeneous networks (Hetnets). Combining all of these technological solutions the peak data rates are considerably improved, as well as cell edge performance and coverage. Carrier aggregation (CA) in LTE-A systems allow the use of up to 100 MHz for a single user while it maintains backward compatibility with previous versions of LTE systems. LTE-A runs on frequencies ranging from 40 to 100 MHz. LTE-A is an upgrade that allows 1 Gbps download and 500 Mbps upload speeds for the network. Relaying is also improved in LTE-A. This is done by having RNs utilized by LTE's eNodeBs and the UE. A final important addition to LTE-A is the concept of a SON. The main purpose of SON in LTE-A is to reduce capital expenditure (CAPEX) and operating expenses (OPEX).

The objective of this chapter is to present a completer review of LTE and discuss several technologies that have been utilized to overcome the staggering increase in data and capacity demands on wireless networks. In addition to our comprehensive discussion on LTE, the simulation results of our study show that LTE-A is truly the 4G technology that meets the ITU requirements.

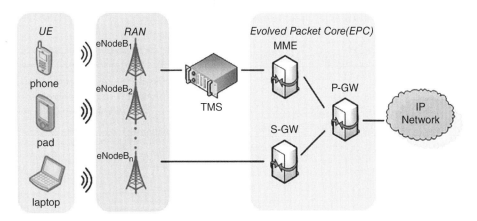

Figure 8.3 The LTE network architecture. (https://www.researchgate.net/publication/319488685_A_Hybrid_Markov-based_Model_for_Human_Mobility_Prediction/figures?lo=1).

8.2 Network Architecture

There are a handful of components and technologies that make up the architecture of the 4G LTE network. They include the device the user actually possesses to access the network and use the service to the gateways that transport the data out across the LTE network to their destinations and back. One advantage seen here is that the LTE network and its switch to all IP based communications is that it is much simpler in design and complexity (so far as the path the data will have to traverses to get where it needs to go) than the third-generation mobile communications system it is now replacing. You can see the basic setup in Figure 8.3. The paragraphs following will delve into each part pictured.

8.3 User Equipment

The first piece of the LTE architecture is something most people carry with them in one form or the other nowadays – user equipment/mobile station (UE/MS). The UE is essentially an application level user interface that signals when and how to setup, maintain, and end transmission initiated by the user. Overall control of the UE within the LTE architecture is handled by the core network.

8.4 eNodeB

The radio access network for 4G has been consolidated from having several different vendor and protocol specific nodes to only have one for the entire network. The network is known as E-UTRAN (Evolved Universal Terrestrial Radio Network). The number of nodes has also decreased allowing for better efficiency and cost savings at implementation and throughout the operation of the nodes. These nodes are all called evolved Node B (eNodeB) and the nodes each handle the LTE signal over a certain geographic area. Each node has the capability of talking to other nodes as well as the UE to move data through the system. The eNodeBs communicate with each other via interfaces called X2 interfaces (and usually just referred to as "X2"). These logical interfaces provide connection not only directly between each node, but also with the EPC (evolved packet core) network (though communication with the EPC it is done with "S1" ports instead). There are several other functions handled by the E-UTRAN and eNodeBs. The eNodeBs handle header compression, reliability of packet delivery, and ciphering. On the control side, eNodeB functions include admission control and radio resource allocation and management. Many of the functions integrated into the eNodeBs were handled by radio network controllers (RNCs) in previous data systems. This is beneficial because the eNodeBs provide less transmission latency with fewer hops as well as better distribution for processing of the load.

8.5 Radio Access Network

C-RAN is an eRAN (evolved radio access network technology that aims to overcome the limitations of legacy RAN systems. The current RAN systems were designed a bit shortsighted. They are all built on rigid and inflexible proprietary bases and present with a few major issues. First is that each base station (BS) can only connect to a certain, non-modifiable number of antennas covering a relatively small service area. Second, is that they are very susceptible to interference. Lastly, there is still a huge amount power required to run each of these units. This not only runs counter to modern thoughts on green operations, but the cost involved is of course a great source of cost for mobile telecommunications providers (Kullin and Ran 2011). Interconnectivity between towers will be a key feature in eRANs including cloud-RAN. C-RAN will attempt to simplify and save money largely by pooling baseband processing at a single, larger location (that can facilitate communications between other such setups). Simplification comes from using a single type of node in the unified station itself where before many were used. These are the new LTE eNodeBs. An additional benefit comes from having the baseband processing in a centralized location when taking into consideration the ability to have a small number of technicians able to support many systems from one central location versus many spread out.

8.5.1 Gateways and Mobility Management Entities

The mobile management entity (MME) we mentioned above is solely used for signaling. User generated IP packets do not pass through the MME. The MME provides a separate network component for signaling to allow network operators to increase signaling and traffic capacity on the network independently. The MME performs a specific function in user authentication/authorization, user tracking, security functions, and network architecture specific (NAS) signaling. NAS signaling occurs to and from the actual UE. The primary job of the MME is to manage and ensure user mobility (Blom et al. 2010).

The first of two gateway systems involved with LTE is the serving gateway (S-GW). This gateway's primary function lies with tunneling of data connections from the UE. It forwards the packets through the LTE network (Siddiqui, Zeadally, and Fowler). It acts as an anchor for the UE when connecting between legacy 2G/3G networks as well as other networks when they go through the S-GW into the packet data network gateway (PDN-GW), which will be discussed next. The MME can also access the S-GW to request tunneling of data between eNodeBs. In these cases, the MME will request and terminate tunnels as necessary. In the case of the UE utilizing the S-GW, the SG-W transfers data on active connections from the eNodeB to the PDN-GW. UEs can however go into idle mode when not using their data connection, which releases the allocation they were utilizing to other devices. When the S-GW receives packets destined for the collapsed tunnel it buffers them and has the MME begin paging the intended UE. This paging process will force the UE's tunnel to be reconnected and the packets that were buffered by the SG-W to be forward on to the receiving UE.

The second gateway that is part of the core LTE network is the PDN-GW. The PDN-GW acts as the external router for the LTE network to other IP networks, the internet in particular; it is a bridge between the EPS (evolved packet system) and those external networks. A UE is able to connect to multiple packet data networks (PDNs) at once to facilitate connections to multiple different PDNs. These gateways are located and maintained at the provider/operator's premises. They also perform filtering functions as necessary. The PDN-GW is also responsible for dynamic host control protocol (DHCP) services as well. When the UE connects to other PDNs it must have an IP address to communicate. The PDN-GW will provide an IP each time the UE requests a new PDN connection or when a when new connection to a different PDN is necessary. The PDN-GW can also perform accounting functions on the data passing through it.

8.6 Evolved Packet Core (EPC)

Figure 8.4 shows a schematic diagram of a legacy 2G/3G and a LTE. The EPC is a component of the EPS. The EPS is an all new packet system derived for the LTE system. The main responsibility and feature of the EPC is "to provide the IP-layer seamless mobility when a UE moves between different eNodeBs (Chen et al. 2011)." Not only does the EPC allow complete IP data communication for the new LTE network, but it springs into the future where many "current" technologies are lagging by implementing a protocol called Proxy Mobile Internet Protocol v6 (PMIPv6). This allows seamless use of the IPv6 network versus the outgoing IPv4 network. The creation and implementation of PMIPv6 involved cooperation between 3GPP and the IETF (Internet Engineering Task Force). Together they created and honed the standard need for an all-internet protocol network (AIPN). This achieves successful packet flow via PCC (policy charging control). PMIPv6 was made to replace Mobile Internet Protocol v6 (MIP) due to handoff and security issues. PMIPv6 uses a device known as a mobile access gateway (MAG), which is usually a router, to serve as the "proxy" between MNs (mobile nodes [UE]) and networks such as the internet. This shields the user from the possibility of some malicious attacks.

The all-IP based functionality of the EPC means it opens up a world of new possibilities for high-speed truly mobile services. As we briefly discussed earlier, the fact that all data can now travel as IP packets across a truly high-speed network with a much higher capacity, we can now find LTE services in numerous varied situations. Whether it is in a car aiding in a telematics service (or even just delivering Pandora internet radio and up to the second weather information) or providing high-speed service to half a dozen devices

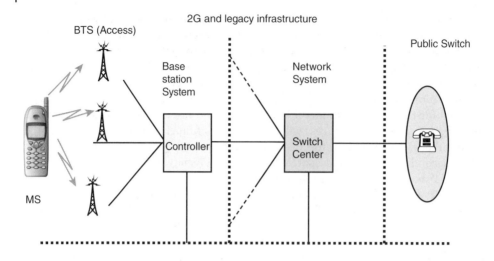

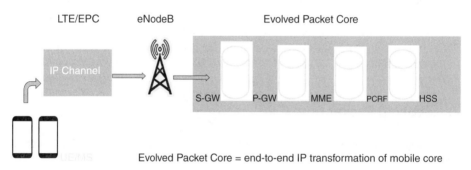

Figure 8.4 LTE/EPC versus legacy mobile data systems (http://next-generation-communications .tmcnet.com/topics/end-to-end-ip-transformation/articles/53890-introduction-evolved-packet-core .htm).

by use of a mobile hotspot that shares the LTE signal via Wi-Fi, LTE means people can work and play on their mobile devices like never before. Even on devices that are not classified as smartphones, both voice and data alike can now move easily across the air with no wires needed – and with superb QoS.

8.7   The Innovative Technologies

LTE, with its more efficient use of spectrum and an impressive road map of features to increase capacity, is a crucial element in how operators will address demand. While LTE is still relatively new, there have been some enhancements made to the system already. LTE utilizes OFDMA to share the available bandwidth of downlink connections. OFDMA shares the available bandwidth using extremely narrow subcarriers. This splitting of the bandwidth allows data to be "chopped up" to just the amount of bandwidth necessary and sent over multiple paths (also utilizing IP) to the receiving end. The smallest time-frequency unit measured for these transmissions is a resource element that equals one symbol on one subcarrier (Siddiqui et al. 2013). OFDMA allows for stronger multi-route resistance through the network (Chang and Pan 2012).

LTE-A runs on frequencies ranging from 40 to 100 MHz (a very wide range) versus 1.25–10 MHz for the original iteration of LTE. These enhancements are all made to be backwards compatible with all earlier LTE standards and are known as LTE-A. LTE-A is the upgrade that allows 1 Gbps download and 500 Mbps upload speeds for the network. There are several other enhancements presented with LTE-A. One is carrier aggregation, which allows combination of nearby frequencies. There is also an enhanced multiple access system known as clustered SC-FDMA (single carrier frequency domain multiple access), also sometimes called discrete Fourier transform spread orthogonal frequency digital multiplexing (DFT-S-OFDM) (Siddiqui et al. 2013). This system allows non-contiguous groups of carriers to be grouped together for a use by a single UE, allowing more efficient link performance. Enhanced multiple antenna transmission allows up to 8×8 spatial multiplexing, which increases downlink spectrum efficiency.

In addition, LTE uses MIMO. MIMO is a form of spatial multiplexing that means that at both the sending and receiving end, multiple antennas in and out are used to transport data across the LTE network to the UE. Much like internet traffic MIMO allows data packets to travel different paths and arrive at their destination at different times using multiple paths. A 2×2 system (2 in, 2 out) could theoretically double the throughput for users. This is very dependent, however, on several deciding factors, the major one being the UE (smart phone, tablet, etc.) utilized and its ability to efficiently use MIMO. Figures 8.5 and 8.6 present the result of the simulation to show how throughput and coverage are affected by MIMO application in LTE.

In Figure 8.5, single layer, 1×1, corresponds to the traditional downlink configuration with one TX and one RX antenna. Single layer, 1×2: one data stream is transmitted, using one antenna, but the receiver has two receive antennas, improving coverage compared to the single layer 1×1 case. Single layer, 2×2: one data stream is transmitted,

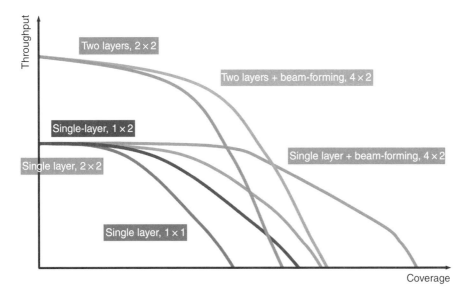

Figure 8.5 Antenna solutions imply different capacity/coverage trade-offs.

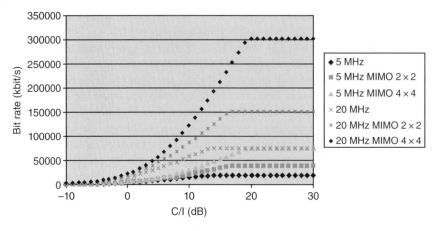

Figure 8.6 Bit rate versus carrier to interference ratio (C/I).

also using two antennas for transmission, which further improves coverage. Two layers, 2 × 2: two data streams are transmitted, potentially doubling the bit rate in areas with good carrier to interference (C/I) (low path loss) but providing poorer coverage in areas with poor C/I. In practice, we can switch to single layer 2 × 2 at the point where the two curves intersect in order to increase robustness in difficult radio conditions. Single layer + beam-forming, 4 × 2: a single data stream is transmitted, but using *four* TX antennas in order to "focus" the transmission. In this way coverage can be substantially extended. Two layers + beam-forming, 4 × 2: two data streams are transmitted, again allowing a doubling of the bit rate if the C/I is good (low path loss as shown in Figure 8.6). In practice, we may combine this configuration with the preceding one, thus operating at the envelope of the two curves and switching between the two configurations as dictated by the radio conditions. Compare the pairing of single layer 2 × 2 and two layers 2 × 2 discussed above.

Some other enhancements in LTE-A include CoMP and support for heterogeneous networks. CoMP is intended to improve cell coverage and efficiency. It works when data is in the cell edge (edge of nodes transmission where they meet the edge of other node's transmission range) and therefore can be split between different nodes and paths on the downlink and uplink. The CoMP coordinates these signals together to improve efficiency and conduct interference avoidance. Two technologies are used to achieve this. Coordinated scheduling/coordinated beamforcing (CBF) aims to improve performance by having the signal sent from a serving cell tower to the UE while the coordinated scheduling works with nearby cells to control other cells' transmission as to reduce interference of varying signals between the different cells. CoMP functions for both homogenous and heterogeneous networks. joint processing/joint transmission (JP/JT) works by actually having the signal intended for an EU transmitted across multiple cell sites. The transmission is then brought down into a single transmitter that has antennas that are geographically separated.

Relaying is also improved in LTE-A. This is done by having RNs utilized by LTE eNodeBs and the UE. It does this by enabling traffic signaling and traffic forwarding between the nodes and UE to improve high data rates, overall cell coverage, and improve cell edge coverage. A node called a donor node connects the RNs to the eNodeBs.

A final important addition to LTE-A is the concept of a SON. The main purpose of a SON in LTE-A is to reduce capital expenditure (CAPEX) and operating expenses (OPEX). The focus of this with LTE-A involves thoughts toward network growth and optimizing capacity/coverage in heterogeneous 2G/3G radio environments. Additional enhancements are still being considered for LTE to lower OPEX as well.

8.8 PS-LTE and Public Safety

Currently, most public safety agencies (PSAs) (i.e. police, fire, and other emergency response agencies) use specialized communication systems based on narrow band radio network systems. These are the commonly observed "radios", which transmit voice data between police and fire personnel. These systems are the tried and true, but are limited in scope. They provide voice communication, but any requirements beyond that require a secondary system, which creates a problem. Requiring PSA personnel to carry multiple devices is not reasonable, therefore it is preferable to establish network medium that can not only provide MC-PTT (mission-critical push-to-talk), but can also leverage existing technologies such as data and video transmission. The existing technology most likely to support this need is LTE. LTE, as anyone who has used a cellular telephone is probably aware, is the name given to current cellular/mobile band network data communications. Making use of the existing infrastructure to carry both MC-PTT and other mission-critical applications, such as video and data, makes a lot of sense in regards to the unnecessary recreation of the wheel.

LTE refers to the standard for mobile commercials communication. The key characteristics of LTE technology are the few network elements also known as flat architecture. This feature tags along with superior spectrum utility and huge broadband capabilities as compared to radio communications. LTE is endorsed by public safety organizations including NEMA and APCO with their 700 MHz public safety radio band preferred in the United States (Anritus, 2019). The safety communication system is relied upon by first responders as it allows for interactive sharing, real-time video links of crime scenes and disaster hit areas from multiple angles, enabling quick analysis and evaluation thus enabling quick reaction. Several worldwide network carriers have announced plans to convert their networks to LTE including Verizon Wireless and AT&T Mobility (Dano, 2017). Much of this technology is characterized by adopting 4G and 5G services for commercial use. The public and safety communication is guaranteed through reliance on cell communication in an emergency situation.

With LTE ready for the full portfolio of services, prioritization of traffic classes needs to be done to ensure mission-critical users have unlimited access to the push-to-talk, video and data services enabling traffic focus on important services by users. This has raised eyebrows as it has been asserted that a compromise for the end users due to overflow can result in traffic congestion that can lead to large network exchange on densely populated areas and in sporting, music and emergency services

PS-LTE stands for public safety powered by LTE. According to Samsung, "One of the biggest advantages of public safety-Long Term Evolution (PS-LTE) networks is data connectivity. While previous public safety networks could only relay voice, PS-LTE can transfer high volumes of data instantly" (2017). Overall society will greatly benefit from the use of PS-LTE. Emergency situations will be projected more quickly, will be more easily tracked, and easier contact will be made when PS-LTE is being used. The number of emergencies and disasters is very large; however, with the use of PS-LTE, the communication during these disasters or emergencies will become easier. This will be done through the combination of different safety networks, which is more easily done with the use of LTE in relation to public safety.

8.9 PS-LTE

According to GSA, "To respond adequately to disaster situations and to ensure public safety and security, first responders need the latest communications technologies" (2018a,b). PS-LTE is a necessary thing in today's time. With the world of telecommunications constantly growing and changing, there is no reason why the world of public safety in relation to LTE cannot be growing and changing as well. Without the use of LTE in the field of public safety, there may be injuries or deaths occurring that could possibly be prevented if LTE was being used.

The Global Mobile Suppliers Association (GSA) has revealed some statistics that show the growth of LTE in relation to public safety. According to their research, "10 countries are deploying or have launched PS-LTE networks, 9 countries have plans or trials in place for PS-LTE networks, 78 different devices have been designed for PS-LTE networks, 42 different types of smartphones or other handheld devices support PS-LTE and finally there are approximately 35 different suppliers that provide PS-LTE devices" (GSA 2018). This shows how great the growth of PS-LTE really is and how it is becoming a large part of telecommunications around the world.

8.10 Nationwide Public Safety Communication Systems

Before discussing how public safety and LTE should work together, it is first important to explain why there is a desire for this. Across the country, there is a growing desire for a nationwide public safety communication system.

8.11 Advantages of LTE Technology

High coverage is another advantage set to make indoor activities more powerful as enhancements in building penetration will help implement 4G Lite with a higher frequency band. Verizon Wireless' deployment of LTE provides coverage within a specific range within a premier 700 MHz spectrum. This enhances in-house productivity over the previous versions. LTE is also set to offer inherent support to IP thus ensuring mass coverage. With this, the need for public safety and technology co-relation is backed by public safety organizations, making it easy to adopt LTE technology. Real-time access to mission correlated information anytime anywhere helps coordinating and responding quickly and thus saves lives. In a single multi-purpose handset, the modernized streamlining and communication service access has reduced the cost of maintaining multiple non-interoperable groups of networks.

There is a significant lack of LTE services, especially in the case of an emergency, in areas that are rural. It is a necessity for there to be a way for emergency news to reach more rural areas around the country, or even to be able to reach anyone no matter where they are in the country. In order for this desire to be met, there must be the development of a nationwide public safety communication system.

Another one of the main reasons that a nationwide public safety communication system is desired by officials that work in public safety, is because they desire a very reliable way to protect the public and inform everyone at the same time if there is an emergency or disaster. With the use of LTE in relation to public safety, this will develop a much more reliable way for those who work in public safety to communicate with all of the public, no matter when or where the disaster or incident may occur. Without question, communication reliability toward public safety must be affirmative especially under stressed conditions including loss of power, lack of operational personnel and loss of infrastructure. This calls for the need of self-supporting centers that act as an emergency regional resilience center. With broadband services, there is a demand for improved safety as terrorist activities and responders find it also critical to have their safety guaranteed during operations. It is thus of essence to ensure the use of LTE technology guarantees safety across the different departments and the whole population.

Reduction of cost is another advantage that has pressurized the implementation of public safety networks as cost control needs to be managed at the same time improving communication across the different channels.

Higher data rates with up to 5–12 Mbps in the forward link, and 2–5 Mbps in the reverse link, LTE will be supporting average data rates per user. LTE video application is set to be enhanced with downlink as well as uplink that is restricted to be within the 3G restrictions, which are the same as of today due to the fact that both maximum and average LTE data rates are significantly higher in reverse and forward links than supported by 3G technology.

Finally, one of the other biggest reasons that a nationwide public safety communication system is desired is because it will provide a more stable way for officials from different areas of the public safety field to communicate with each other. Public safety officers will be able to communicate with other public safety officers across the country when needed and will not have to worry about other officials being misinformed or not being able to reach or contact each other in case of an emergency or disaster. Figure 8.7 shows a transformation of technologies in public safety communications systems.

8.12 Driving Trends in Public Safety Communications

There are also many trending topics in relation to the world of communication in public safety. Convergence of devices, networks and services for the purpose of interoperability and reliability is the main issue that needs to be solved. Figure 8.8 shows a possible network integration of private and public systems where all public safety entities can communicate through an integrated platform with the public. Todd Piett mentions three main trends that are occurring in the years of 2017 and 2018 in regards to public safety and communication. These three trends include; "crowdsourcing and

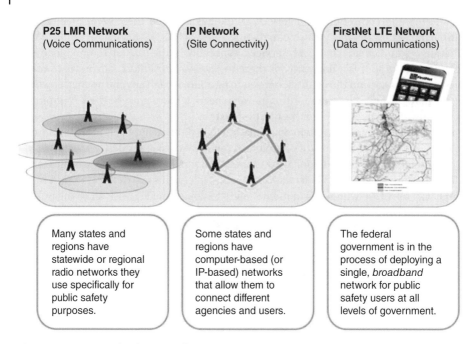

P25 LMR Network (Voice Communications)	IP Network (Site Connectivity)	FirstNet LTE Network (Data Communications)
Many states and regions have statewide or regional radio networks they use specifically for public safety purposes.	Some states and regions have computer-based (or IP-based) networks that allow them to connect different agencies and users.	The federal government is in the process of deploying a single, *broadband* network for public safety users at all levels of government.

Figure 8.7 Various technologies used in an emergency situation. https://slideplayer.com/slide/11338308.

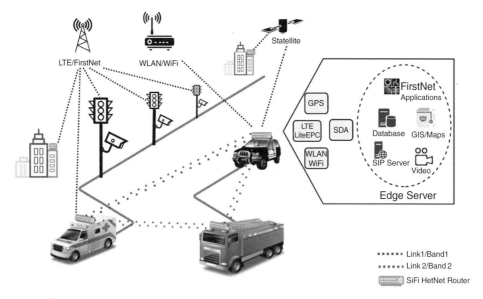

Figure 8.8 Wireless communications systems for public safety. http://www.spectronn.com/public-safety-communications/resilient-wireless-networking-for-public-safety-combining-hetnet-fog-computing-and-dynamic-spectrum-management.

engagement, private and rapid communication, and improved collaboration across silos" (Piett 2017).

The first of these trends that is crowdsourcing and engagement. Often people will want to report if they are a witness to a crime, or even to report a crime that has happened to them. However, there is not always a clear and confidential way to report the crimes that they have been present for. This is one of the things that is trending and that people across the country want a response to. There should be a secure line or some secure form of communication between the public and safety officials where they feel safe to communicate with them without being afraid of hackers, people over listening, etc.

The second trending topic is in regards to private and rapid communication is that one of the biggest issues in the world of public safety is the lack of secure and private communication. There has to be security and protection for the information that is being passed over to PSA and officials. Especially in the case of emergencies, the release of secure information can cause chaos and can even make the situation worse if the confidential information regarding the emergency is not kept confident and secure. For this reason, there has to be a more stable and secure line of communication between different public safety and law enforcement agencies to protect the public.

Finally, there is the trending topic that regards improved collaboration across silos. This is similarly related to the topic of the nationwide public safety communication system. There is a great need for there to be a system put into place that will allow public safety officials to communicate directly with the people, no matter where or what time an emergency or disaster may occur. This is a trending topic due to the importance and relevance it has to our current society at this time. There should be a stable and secure system that allows residents in the United States of America to learn of an emergency or national natural disaster that is quick and efficient. We should not have to go searching for information in this generation with the considerable amount of technology that we have, the information should come to us quickly and all should receive the information.

8.13 Benefits of PS-LTE

Next, we will discuss the benefits of PS-LTE. The first main benefit of having PS-LTE is that there is a wireless communications network that has larger access than ever before. However, there are also many other benefits. One of these main benefits is "First responders will have full access to broadband data, allowing the introduction of new mobile applications for everything from evidence collection to e-citations and SOS alerts that notify dispatch of the location of an officer via GPS. Persistent access to record management, databases, and smart analytics will lead to more efficient policing, coordinated life-saving efforts and crime solving" (Jones 2018).

At the moment, the use of LTE is not very present in the public safety world. The most common form of communication between law enforcement or public safety officials is through standard landlines. This limits the amount of communication that can occur between officials over the different departments of law enforcement and public safety. This also limits where public safety workers may have to be in order to receive important information or in order to communicate with other branches of public safety.

This greatly limits the amount of communication and causes the amount of time that it takes for officials and public safety workers to be informed or notified of emergencies or disasters.

One of the most current and useful attempts at PS-LTE is the introduction of FirstNet to our nation. FirstNet stands for the First Responder Network Authority. FirstNet is described as: "FirstNet is an independent authority within the US Department of Commerce. Authorized by Congress in 2012, its mission is to develop, build and operate the nationwide, broadband network that equips first responders to save lives and protect US communities" (First Responder, 2018).

According to the CEO of FirstNet, Mike Poth, "One of the key benefits that this public-private partnership will make available to public safety is the availability of quality of service and priority access once a governor accepts the FirstNet State Plan ('opt-in'), with preemption services expected to be made available on all AT&T LTE bands as soon as the end of this year," (Poth 2017).

This will have a big influence on the relationship between public safety and LTE. This system will allow for public safety officials to notify everyone around the country, no matter where they might be, about any natural disaster or emergency that has occurred in their area, or even nationwide. This system is a new form of technology that has not been used nationwide in the United States of America before. It is a great step in the right direction for the world of PS-LTE and is a great example of what the PS-LTE world can do, and how it can change the way that the nation and the world look at public safety in the eye of technology.

It is time that technology is used in a way in which we can all benefit from, and the use of the FIrstNet system will provide us all with more safety and comfort, as well as making the jobs of law enforcement officials and public safety officers much easier. Without the use of this technology in previous years, it has been an issue, and it has not been easy to inform those across the country in the case of natural disasters, emergencies, etc.

With the use of this technology, there is no reason why anyone should be without the necessary knowledge regarding their safety or the safety of the state or city that they live in. With the use of this technology, people around the nation will benefit and will be provided with the protection and information needed.

Another benefit of PS-LTE is the ability to transfer data more quickly. The field of public safety often involves the need for the transfer of data between different officials and offices of public safety. This may include messages, emails, pictures, videos, files, etc. For this reason, there has to be a quick and reliable way for public safety officials to be able to send and receive data over a reliable network. With the use of PS-LTE, officials would be able to send all of this necessary information using LTE internet. This allows for officials to receive information and data more quickly and also allows for the public to be able to send necessary information to public safety officials in an easier, faster, and more reliable way.

Ingo Flomer explains a few of the benefits that could come from the use of PS-LTE in today's society and how it could provide assistance to the world of public safety. "Facial recognition capability by simply taking a photo of a member of the public who has been arrested could speed up the process of dealing with them when they have broken the law. The image capturing and uploading in a disaster scenario could provide crucial evidence or intelligence further down the line when the scene has been tampered with following rescue efforts when vital clues might have been lost" (Flomer 2015).

Overall, the use of PS-LTE will allow the public to communicate with public safety officials and will allow public safety officials to communicate with law enforcement officials more effectively and in a more efficient way. This will occur in real time rather than taking a long amount of time to send and receive information. Overall, the whole nation will benefit from the use of PS-LTE and it will hopefully prevent crimes, deaths, and other injuries.

8.14   Benefits of Converged Networking in Public Safety

The next topic that we will discuss is the benefits of converged networking in a public safety environment. Firstly, network convergence can be defined as "the delivery of a range of communications services through a single cable" (Wellinger 2015). Basically, this means that there is a combination of telephone, internet, and security that can be integrated into one pool that will complete the computing, networking necessities, and storage all in one.

One of the main benefits of converged networking in the world of public safety is that the systems overall will be smarter. Convergence makes the system more intelligent and will allow for more efficient data transfer. This will improve the IT resources that are involved in the system and will allow for the systems to be combined into one. This will allow for the combination of internet, telephone, security, and data storage needs in the world of public safety. This is very important due to the large amounts of information that public safety officers receive daily that must be received quickly. It also important to store information in a safe and reliable environments that will securely protect the information that is found by them or sent into them.

Another major benefit of the use of converged networking in relation to public safety is the lower cost. Due to the fact that many areas are being combined into one, the cost will obviously be lower. The cable that is required for this technology is smaller making it easier to store, and will also require fewer servers, switches, etc. This will be the first way that the system will save public safety departments money. Secondly, the combination of the different areas, internet, telephone, and data security and storage, will allow for public safety officials to save costs. There will not be a need to pay for each individual system and there will not be the need for the system to be kept separately from the beginning.

Finally, another way that converged networking in a public safety environment will save money is because of the amount of power saving that it will cause. Due to the fact that the converged networks will be deployed, the overall amount of power that will be consumed will become less. This will allow them to save energy and will lower the cost of electricity, as well as the amount of equipment that will need to be purchased will become less in this case.

8.15   Mobilizing Law Enforcement

Mobilizing law enforcement in relation to PS-LTE is the next topic we will discuss. Dale Stockton states "For decades, public safety personnel have relied on dedicated radio systems to support field operations and ensure the timely sharing of mission-critical information. When radio frequencies became crowded with traffic, many agencies ventured into cellular communications with the introduction of computer-aided dispatch (CAD) systems supported by mobile computers that relied on data carried across commercial wireless networks." This explains the previous issues that law enforcement and public safety officers may have faced in regards to technology in relation to their jobs.

However, with the new development of PS-LTE and its growth, technology has grown and the standards for communication in the law enforcement and public safety community have grown along with it. The government is more responsible for the safety of the public, so it is necessary that they provide new technological opportunities and help discover new opportunities in order to assist law enforcement officials and public safety officials in the protection of the communities around them. 4G LTE is cost effective, reliable, effective, and allows remote management and connectivity, see Figure 8.9.

Law enforcement and public safety officials are not officials who stay in one place. They are constantly moving and traveling to different areas based on where they are needed. One day a law enforcement official may be needed in a place at a much farther distance from where he or she is used to. For this reason, it is necessary that the technology that is being used by the law enforcement and public safety officials meets the needs of what they will be doing, where they will be going, and wherever they may be needed or may have to travel to.

Communication is the key to public safety and communication it is becoming even more simple with the growth of PS-LTE. Each year the amount of information that law officials receive from the public around them that provides assistance grows and grows. People across the nation will send in information in regards to public safety, events, and other national disasters that may occur.

Currently, due to the growth of communication at events, it is easy for an event to become a dangerous place for many people to be or for a crime to occur. So, for this reason, it is necessary that officers and public safety officials have all of the tools and

Figure 8.9 Benefits of 4G LTE for security and police officers. https://cradlepoint.com/blog/cradlepoint/5-reasons-police-public-safety-embrace-4g-lte (Cradlepoint 2015).

necessities that they need to protect the public from where they are, without suffering due to the lack of the technology that they need to complete their jobs.

While currently law enforcement and public safety officers are able to travel around and communicate with each other no matter where they are working, there are many difficulties that they should not have to face that they are currently facing.

Communication between law enforcement officials and public safety officials should be easy no matter where they are. Anything that a law enforcement official or public safety official can do at his place of work, he should also be able to do on the go. There is no reason why in our current generation, an officer should have to return to his office or work in order to complete a simple task. Viewing important files, pictures, videos, communicating with other law enforcement agencies, or public safety officials should be easier than ever now that we have the technology (PS-LTE) to support these needs and wants of the officers.

One of the most important reasons for the need of PS-LTE and the mobilization of law enforcement officials and public safety officials is due to the growing number of events, concerts, meetings and other public activities that require public safety and law enforcement officials to be present for protection.

These situations are some of the toughest for law enforcement officials to monitor and control due to the lack of communication between the different law enforcement officers and other public safety officials at such events, as well as due to a large number of people that may be present.

If officers are to have all of the tools and necessary technology that they have at their own offices or places of work, then there should be no reason why they should not be able to determine who someone is using facial recognition software through a smartphone or to be able to communicate with other law enforcement agencies if there is an emergency or disaster at or nearby the event.

In conclusion, PS-LTE is a growing aspect in the world of telecommunications and technology today. There is a growing need to protect the people of the United States of America and the need for more communication between law enforcement officials and public safety officers. For these reasons, it is necessary that public safety and LTE work and grow together to provide the best possible protection for the people of America.

References

Anritsu Effectively Testing 700 MHz Public Safety. (2019). Anritsu Retrieved from https://www.anritsu.com/en-us/test-measurement/solutions/en-us/effectively-testing-700mhz-public-safety

Blom, R., Normann, K., Naslund, M. et al. (2010). Security in the evolved packet system. Ericsson Review. *The Communication Technolgy Journal* (2): 4–9.

Chang, Y. (2012). *Computer Science and its Applications*, 217–229. New York, NY: Springer Netherlands.

Chen, Y., Juang, T., and Lin, Y. (2011). A secure relay-assisted handover protocol for proxy mobile ipv6 in 3gpp lte systems. *Wireless Personal Communications* 61 (4): 629–656. https://doi.org/10.1007/s11277-011-0424-2.

Cradlepoint (2015). 5 Reasons Police & Public Safety Embrace 4G LTE. (2015). Retrieved from https://cradlepoint.com/blog/cradlepoint/5-reasons-police-public-safety-embrace-4g-lte

Dano, M.: All AT&T LTE bands to be available to public safety this year. (2017, May 01). Retrieved from https://www.fiercewireless.com/wireless/firstnet-all-at-t-lte-bands-to-be-available-to-public-safety-year

First Responder Network Authority. (2018). FirstNet Retrieved from https://www.firstnet.gov/sites/default/files/PSAC_Charter_20190402.pdf, December 2018

Flomer, I. (2015, July 17). Will LTE ever become a reality in public safety communications networks? - Axell Wireless. Retrieved from https://cobhamwireless.com/resources/blog/will-lte-ever-become-a-reality-in-public-safety-communications-networks

GSA 2018 LTE Critical Comms - Public Safety LTE Networks and Devices – Market Status July 2018 GSA https://gsacom.com/paper/public-safety-lte-networks-and-devices-market-status-july-2018/

Jones, R. (2018, June 25). How PS-LTE Is Bringing First Response to the 21st Century. Retrieved from https://insights.samsung.com/2018/06/12/how-ps-lte-is-bringing-public-safety-communications-into-the-21st-century

Kullin, C., & Ran, D. (2011). C-ran the road towards green ran, White Paper, 2011. China Mobile Research Institute

Lawson, S. (2012, August 07). Growth in new mobile-network software could help shape services, bills. Retrieved from IDG News Service http://www.cio.com/article/713133/Growth_in_New_Mobile_network_Software_Could_Help_Shape_Services_Bills

Piett, T. (2017). Public safety communication trends in 2018 Retrieved from https://www.ravemobilesafety.com/blog/2018-public-safety-communication-trends

Poth, M. (2017). Written Testimony of Chief Executive Officer, First Responder Network Authority (FirstNet). https://docs.house.gov/meetings/IF/IF16/20171101/106569/HHRG-115-IF16-Wstate PothM 20171101-U10.pdf

Public Safety LTE Networks and Devices - Market Status July 2018. (2018a). Retrieved from https://gsacom.com/paper/public-safety-lte-networks-and-devices-market-status-july-2018

Public Safety LTE. (2018b). Druid Software Limited Retrieved from https://www.druidsoftware.com/public-safety-networks

Public Safety LTE (PS-LTE) (2017) Networks Solutions Samsung Business Global. (2017, December 20). Samsung Retrieved from https://www.samsung.com/global/business/networks/solutions/public-safety-lte

Siddiqui, F., Zeadally, S., and Fowler, S. (2013). *Next generation Wireless Technologies*, 71–103. London, UK: Springer-Verlag.

Smith, B. (2010). The shift from 3G to 4G. Wireless Week, Retrieved from http://www.wirelessweek.com/articles/2009/01/shift-3g-4g

Wellinger, T. (2015, August 21). What are the benefits of network convergence? Retrieved from https://thestack.com/world/2014/04/21/what-are-the-benefits-of-network-convergence

9

4G and 5G for PS: Technology Options, Issues, and Challenges

9.1 Introduction

Over the years, there has been evolution in almost all sectors in the world. One such area where there has been an advancement is regarding technology. There have been many forms of improvement and advancement in technology, which means that new technology is better than the previous technology. The changes in technology can either be an improvement or a complete departure from the tenets of the previous technology. One such area where there has been a constant and drastic improvement over the years is in the wireless network where, currently, there is 4G. As seen from the way it is written, 4G refers to the fourth generation of mobile phone technology. As a result of being the fourth generation, it means it had previous generations, specifically 3G, which was third generation mobile phone technology, 2G which was second generation mobile phone technology, and 1G which was first generation mobile technology and the pioneer in mobile phone technology.

As expected, the next phone generation language is better than the first generation language. 1G, which was a first-generation language was an analog kind of technology. It was therefore very limited in all that it could do, with no possibility of making voice calls over the internet. The essential thing it could do was the sending of text messages over the local area's network. In the 1990s, there was an advancement in the generations of phone technology with the launch and implementation of 2G technology. This second generation technology was digital and not analog like the 1G. With the use of 2G, there was a remarkable improvement from the 1G services, hence it was able to send messages both over the local area network as well as through the internet, and it was also now possible to make phone calls over the internet, to any part of the world as long as the receiving device was also 2G enabled.

In early 2000, there was also yet another improvement in phone technology, with a switch to third generation phone networks. As expected, this 3G phone technology was better than the 2G phone technology since it was now able to operate at faster speeds in sending messages over the internet and over local area phone networks (Goggin 2012). Moreover, using 3G networks, the voice calls were now clearer than before. The disturbances in terms of noise, decoding, and encoding of the messages were eliminated. Users who communicated using the 3G spectrum could make their calls with little or no interruption in terms of clarity of the phone calls. Moreover, there was also the speed of browsing the internet, with download speeds of between 550 Kbps and 1.5 Mbps. The high-speed data services offered when downloading and low buffering therefore made 3G more favorable than 2G. In 2008 there was another improvement in the phone network generation. 4G was launched and it was a massive improvement on the 3G technology (Mariappan et al. 2016). One of the things that was an improvement was the speed of download. The new speeds averaged 100 Mbps for high mobility communications and 1 Gbps low mobile communications such as stationary and pedestrian users.

9.2   4G LTE and Public Safety Implementation

4G Long Term Evolution (LTE) is an improvement on the basic 4G technology. The reason for the changeof the name is because of the un-attainment of the 4G standards that were set in 2008 by the International Telecommunications Union radio communications sector (Tsai et al. 2017). These standards set were seen at the time to be unachievable, and some have yet to be achieved. These standards included, for mobile use including smartphones and tablets, connection speeds peaking at at least 100 Mbps, and for more stationary uses such as mobile hotspots, at least 1 Gbps. Unfortunately, such speeds have not been met, although, since then, there have been some phone technology producers who have claimed that they offer 4G services. Some have in fact been challenged in court successfully. One such company phone technology manufacturer that has been challenged successfully is Apple. In 2010, it manufactured its products and placed the 4G signs on them. This was in relation to the tests that had been done in the US where there were areas where 4G was enabled. Unfortunately, when Apple sold these phones and other electronics abroad, there were areas where it did not function because the local networks and frequencies in the areas did not have 4G capabilities.

One country that sued Apple was Australia where 4G technologies had not yet been launched and offered at the time. Apple challenged but lost the case as it was determined that it was false advertising (Tiwari et al. 2016). For all the products which the company took internationally, it was then decided that the company had to take out all the 4G signs that it had plastered them in. Another reason why there was a progression to 4G LTE was that the speeds that were touted and set as standards have never be achieved. Though some of them claim to offer the high rates that were proposed, the reality is often different as many factors make the delivery of these speeds impossible.

It has therefore been deemed as realistic to call it 4G LTE. LTE is known as Long-Term Evolution. This means that there is an improvement over time but the speeds that were set previously have not been achieved yet. There has been, however, slow but sure progression over time. 4G LTE has many benefits and should be implemented for public safety. 4G shapes public safety services in the following ways.

9.2.1   Reliability

4G has the advantage that it is reliable, which is why it is being implemented in police departments. This is because police departments need a method of communication that is reliable and is not prone to interruptions (Bye et al. 2015). As a result, IT departments in various police departments prefer it because it is reliable and enables smooth and efficient communications between the officers on the ground and the officers manning the central headquarters. The routing of vehicles has also been made easy with 4G LTE technology. This is because it is smooth and consistent when connected to Wi-Fi, and this is what is needed in police departments. The Wi-Fi that is sourced from the 4G LTE mobile technology means that it can be connected to tablets, body cameras, dashboard cameras, and so on, hence enabling real-time monitoring without interruptions or outages. This is a better way of communicating with the central station and of transferring files.

This reliability comes in many forms, but there is an agreement across the board that is better than the former methods that were used before that involved the plugging in of UDB air cards into laptops by the police when they are in the field. In this case, apart from the above mentioned reliability, there is a risk that when a USB is plugged into the laptops, there may be some delay, or the network may not work as expected. 4G LTE, therefore, come with some form of reliability that should not be underrated at all.

9.2.2 Cost Effectiveness

As with anything, there is always the consideration of cost. It becomes especially important when spending taxpayers' money is involved (Elayoubi and Roberts 2015). Sometimes, even when a high number of resources is being used, it is essential that the service that is purchased is reliable and available without interruption. In the case of 4G LTE and its implementation for the sake of the public, cost effectiveness is also essential. The initial purchase and deployment of 4G LTE is always expensive. This cost is, however, expected to be recouped by the reliability of the service.

In all public initiatives that have had 4G services implemented, they have always been hailed as being cost effective. This is because, as discussed above, it is a reliable technology for the sake of communication. Lu et al. (2016) argue that this reliability means that there is no need to have a backup system, which still requires more funding. This was the case when public services like the police used 3G and 2G technology. This technology was prone to outages and was therefore not reliable. Police stations that used these services had to personally go to the police stations to file reports instead of sending the reports and files from anywhere in the field. In the end, the police officers spent less time on patrolling and keeping order. With the implementation of 4G LTE, however, there has been an increase in cost effectiveness because the officers can file their reports from anywhere in the field. As a result, they usually do not have to go personally to the station, hence enabling them to spend more time in the field and providing more safety to the public. As a result, this has given the public more value for their money.

According to Fedrizzi et al. (2016), one case where 4G has enabled the implementation of public safety is the implementation of cradle point by police departments in California. This service has been hailed because it saves more time both for the police officers as well as for the IT departments. The technology has been installed and activated in the vehicles of these officers. This, therefore, means that once an officer gets into their vehicle, with the router secured in their trunk, they know that they are connected. This is a departure from the past whereby the officers had a modem which they had to plug in to get connected. Unfortunately, when they needed to be connected, and they go into their vehicles, they sometimes found it broken or missing. 4G LTE has eliminated these frustrations and problems of the past, and has meant that public safety has been improved.

9.2.3 Real-Time Communication

Samba (2014) argues that when it comes to the safety of the public, it is essential to have real-time reactions. Such real-time responses sometimes involve communication through some network, which should be reliable and fast. The use of 4G LTE has been significant, and its application and implementation are on the rise, especially with regards to public safety and the police. Police departments are sometimes faced with situations whereby their swift and fast reaction is necessary; this may include reaction and response to an emergency where there is a security threat, the location of a GPS of a vehicle, and so on. In the past, the police had a problem in that they could not respond promptly to some situations because the Wi-Fi and phone networks they were using were slow.

With the application of 4G LTE, the safety of the public has been improved. This is because 4G LTE has enabled the police department to have communication channels that are not only reliable but also fast and advanced. The tracking of a GPRS of a particular vehicle has therefore been easy, and so has the making of phone calls over the internet, hence increasing the response time. Moreover, 4G LTE has also enabled the police to send some forms of information like reports, videos, pictures, and so on that are compressed or not. The general safety and security of the public have therefore been improved.

Apart from the police, the general public is also in need of fast internet services and the making of phone calls over the internet that is fast and reliable. The members of the public are sometimes in danger, and the only way they can get themselves out of this danger is if they can communicate as quickly as possible over a fast and reliable network. 4G LTE has facilitated this and members of the public can send information and data out as soon as possible, whether in the form of messages, documents, reposts, or even phone calls. The members of the public are therefore better off concerning safety.

9.2.4 Remote Deployment and Configuration

Remote implementation and configuration with zero touch have been efficient in the updating of firmware. This enables additional savings for taxpayers and the police, which can allow the deployment of resources elsewhere for the improvements for the sake of the public (Jordan 2017). The key to this is the features in 4G LTE that have enabled cloud computing. When firmware has been implemented in the cloud, it becomes more accessible for sharing it to remote locations with 4G LTE capability. This was not the case in the past whereby 3G and 2G technologies had not the capability of malware computing and deployment. Remarkably, the deployment of the firmware, with the help of 4G LTE enabled cloud computing, has made the installation take approximately five minutes. If such vehicles belonged to security officers, the army or other security enforcement agencies, then they would be able to go back to their duties in a record five minutes. Such a disruption is minimal, and it helps maintain the security of the public.

9.2.5 Flexibility

As with anything in life, flexibility is always crucial. This is one of the features of 4G LTE that was not present in previous generations of phone and network technologies.

According to Behera et al. (2015), 4G LTE enables the scaling up or scaling down of information as necessary. More or less information can, therefore, be fed into the network and retrieved at high speeds, and from remote locations. This feature of the 4G LTE has been essential in many aspects that improve the safety of the public. This includes the fire department, the police, and emergency services. All the information that these teams needs can be accessed from a remote location at high speeds, and there is a very high threshold to the maximum amount of data, information, reports of calls that can be made or received and retrieved at a time. Moreover, 4G LTE enables for the sharing of information and data between the various departments mentioned above. This allows for the synergy of the department, hence maximizing public safety. A good example is when there is a fire in a school. 4G LTE works in such a way that the information can be shared by the police, emergency departments from hospitals, and firefighters concurrently. Their response will, therefore, be coordinated, and they will arrive within the fastest time possible. This will enable the police to investigate and authenticate the cause of the fire, the fire department will be working hard to put out the fire, and the hospital emergency services will be working on treating any casualties. This flexibility of 4G LTE has, therefore, meant that the flexibility has increased and improved the safety of the public.

9.3 Starting Public Safety Implementation Versus Waiting for 5G

According to Schneider and Horn (2015), despite the presence of 4G LTE, there have been promises from the providers of 4G LTE phone technologies that there are plans in place to introduce 5G. 5G, as the name suggests, is fifth-generation phone technology, which is the next step after 4G LTE. From the early press releases and discussions with an expert in the phone technology industry, it has been hailed as being a massive improvement on 4G LTE. Specifically, 5G has been touted to have speeds that are 100 times faster than the currently available 4G networks. Moreover, it has been said that it will create an Internet of Things. The Internet of Things is envisaged as a kind of network whereby many gadgets and equipment that can use networks can be interconnected together. This may vary from the kitchen appliances, sitting room gadgets, office gadgets, and so on. This will create a seamless kind of symbiotic relationship whereby they will work in sync. Moreover, human beings will also be able to have remote access to these devices from wherever they are and manipulate them. All this is meant to improve human experience with these devices and maximize their efficiency.

Despite all these promises about 5G networks, there is, however, one significant disadvantage that is associated with it: it does not yet exist. Back in 2014, industry experts, companies and governments expected that by 2020, the 5G network would already be in use (Schiller et al. 2017). Unfortunately, it currently appears that this timeline will not be met because there has not been any meaningful progress. There have been calls for the government to widen the legally available spectrum. The widening of this spectrum from where it currently is, it is argued, would let the companies shift their wireless to less crowded frequencies that are of higher ranges. Even if this is done, current evidence does not suggest that it would help. For this reason, there have been joint efforts by various technology providers as well as governments to join resources, pool together manpower, and so on, to achieve 5G technologies.

From the releases from experts as well as the companies that are researching these 5G networks, they will be massively superior to the current 4G networks. As a result, 5G would, therefore, be naturally better regarding public safety implementation, as it has excellent features and characteristics compared to 4G networks. Conventional wisdom, however, advises against waiting for 5G before starting public safety implementation. One of the reasons why it should not be expected is that no-one is even sure when 5G will be successfully created and launched. As a result, the wait could be further away than 2020 that had been earlier predicted.

According to Bader et al. (2017), another reason for not waiting for 5G before implementing public safety is because it is a matter of priority that implementation should not be postponed. In any case, the current 4G networks available have capabilities that have been proven all over the world that are useful as far as public safety is concerned. Some examples include the above mentioned examples of the implementation of 4G LTE for use by law enforcement officers in California. 4G LTE is reliable, flexible, fast, and cost effective. As a result, it has made the public safer because police coordination has improved as has the response to matters of public safety.

From the implementation of police departments, it is clear that 4G LTE can be implemented in other spheres to do with public safety and achieve huge success. In the end, therefore, many department and sectors to do with public safety should invest in the implementation of 4G LTE for public safety. Governments should especially do this to promote public safety. 4G LTE can be implemented in the security sector to ensure the public is safe, in the fire department to ensure that the fire services get reports and react promptly to incidences of fires. It can also be used for hospital emergency services. This would be by installing in ambulances a router enabled with 4G LTE networks. The hospital IT teams can then direct the vehicle by giving it GPS signals and the exact locations of the place they should reach immediately. Furthermore, this router and 4G LTE given to the emergency services also helps them determine the best route to the place, by liaising with satellite services. This real-time relay helps the emergency services avoid traffic, bad roads, dangerous roads, and so on.

From the above discussion, it is therefore important to note that waiting for 5G before public safety implementation is unnecessary. This is despite that fact that it will be a massive improvement on all the above mentioned advantages of the 4G LTE network. In addition, there is no concrete timeline on when 5G technology will be available for use. Additionally, the use of the current 4G LTE technologies will make the transition to 5G easier. This is because the current discussions show that 4G LTE and 5G will not be massively different. It will just be an improvement on 4G LTE. The user interface will be the same; it will be easy to transition from the current 4G LTE to 5G when it is eventually invested in, refined and rolled out.

9.4 5G Versus 4G Public Safety Services

All the discussions above agree that 5G is better than 4G. However, will it really be better in terms of public safety? The answer to this is an affirmative yes. The key advantage of 5G, which makes it better than 4G, is speed. It has been predicted that once its development is complete, 5G will be almost 100 times faster than 4G. The key reason for this is because 5G will be using a spectrum that is on higher frequencies, hence there will be will be less crowding and traffic. The sending and the receiving of signals will, therefore, be faster. Public safety will, therefore, be massively improved by using 5G in the following ways.

9.4.1 Video Surveillance

Kaleem et al. (2018) posit that video surveillance is an important tool in any part of society. A video camera is put on highways, back alley streets, supermarkets, hotels, homes, schools, and so on. They are important in the recording of incidences and relaying them to set locations like police departments, school IT departments, and so on. It is meant to deter crime or enable it to be acted on immediately because video surveillance is relayed immediately. There are, however, some areas where the network is not 4G enabled, hence the surveillance cannot be streamed in real time. The surveillance has to be collected in person, which then beats the hope of safety where a prompt response is key. In areas not covered by 4G and local area networks, the surveillance cameras could be 5G enabled. 5G would then use Wi-Fi to provide for the real-time streaming of the surveillance footage, enabling authorities and the public to react as necessary.

9.4.2 Computer-Driven Augmented Reality (AR) Helmet

According to Merwaday and Guvenc (2015), it is envisioned that the authorities will have a computer-driven AR helmet once 5G has been developed. This helmet will have the capability of scanning an area and perceiving any threats that it may receive. These threats will then be processed and sent to the relevant authorities in real time.

From the above discussion, it is therefore important to note that the most important feature of the 5G network that will enable it to improve public safety is speed. By being 100 times faster than 4G, 5G communication will reach the intended target faster, and it can be communicated in areas where normal networks are not available because of its Wi-Fi capability.

9.5 How 5G Will Shape Emergency Services

Emergency services are diverse but the most common is a police emergency, hospital emergency, and fire emergency. For emergency services to be most effective, time is key. Hence there is a need to arrive at the intended destination in the fastest time possible. 5G is predicted to have the capability to help emergency services to arrive at locations faster (Rao and Prasad 2018). Specifically, 5G could be fitted into emergency cars, connected with aerial satellite systems and GPRS locators and have a team working at the headquarters. In this situation, the 5G will helps in the coordination of all the above situations to give the best route for the emergency series to follow. This route that is chosen should be devoid of traffic, bad roads, security threats, and so on. Although this capability is possible at the moment, the speeds are not fast, hence it is inefficient. With 5G predicted to be 100 times faster than 4G, emergency services will get real-time and accurate information, hence enabling them to pass through the routes that will get them to the location of emergency as fast and as safe as possible.

9.6 4G LTE Defined Public Safety Content in 5G

Dahlman et al. (2016) point out that carrier aggregation is the most important feature of 4G LTE that should be included in 5G. Carrier aggregation refers to the combining of many channels from different spectrums and transmitting them simultaneously. Currently, 4G is able to carry around 20 channels from different spectrums at the same time. This has enabled easy syncing and communication between those administrations, sectors, and department working in the different spectrums. 5G should carry on with this important feature and improve on it. Even more channels from difference spectrums could be combined and transmitted simultaneously, hence allowing for even more cooperation and syncing between different departments. This would allow for the sharing of more information, data, calls, and so on to do with public safety. This increased cooperation will eventually improve public safety.

9.7 The Linkage Between 4G–5G Evolution and the Spectrum for Public Safety

Fantacci et al. (2016) posit that the linkage between 4G–5G evolution and public safety is that the 4G spectrum will have to pave the way for the 5G spectrum. This is because if higher frequencies are to be used, they will be expensive and will be available only to a few. These higher 5G frequencies will be auctioned off, and hence will be sold to the highest bidder. It has been predicted that the price could go as high as $7 billion, which is a high and unachievable figure for most providers. Moreover, after it has been sold, it will only be available for a few, which means the masses will not enjoy the public safety that comes with it. As a result, it has been seen that, for now, the best way will be to reconfigure the current 4G spectrum like the 3.6 GHz band and turn it into a 5G spectrum. As such, 5G will be easily and widely available. Hence the public safety advantages that come with it will be enjoyed by many.

9.8 Conclusion

From the above discussion, it is clear to see that network technologies have been changing over the years. These changes have been toward improvement. Following this, the changes over the years have been from 1G up to 4G currently in use. There are efforts to come up with 5G, but the efforts so far have not been successful. It is however hoped that by 2020, 5G will be in use. The use of 4G for public safety has been implemented in some cases, with discussions on whether 5G should be waited for before full implementation for public safety. The common consensus, however, is that the current 4G technologies should be used to implement measures while we wait for 5G.

References

Bader, F., Martinod, L., Baldini, G. et al. (2017). Future evolution of public safety communications in the 5G Era. *Transactions on Emerging Telecommunications Technologies* 28 (3): e3101.

Behera, S.K., Behera, L.K., Madhusudana, C., and Padhi, P. (2015). Spectrum scheduling for 3G and 4G networks by intra-operator flexibility. In: *Computing, Communication & Automation (ICCCA), International Conference on 2015*, 1391–1394. IEEE.

Bye, S. J., Paczkowski, L. W., Parsel, W. M., Schlesener, M. C., & Shipley, T. D. (2015). *U.S. Patent No. 9,161,227*. Washington, DC: U.S. Patent and Trademark Office.

Dahlman, E., Parkvall, S., and Skold, J. (2016). *4G, LTE-advanced Pro and the Road to 5G*. Academic Press.

Elayoubi, S.E. and Roberts, J. (2015, September). Performance and cost-effectiveness of caching in mobile access networks. In: *Proceedings of the 2nd ACM Conference on Information-Centric Networking*, 79–88. ACM.

Fantacci, R., Gei, F., Marabissi, D., and Micciullo, L. (2016). Public safety networks evolution toward broadband: sharing infrastructures and spectrum with commercial systems. *IEEE Communications Magazine* 54 (4): 24–30.

Fedrizzi, R., Goratti, L., Rasheed, T., and Kandeepan, S. (2016). A heuristic approach to mobility robustness in 4G LTE public safety networks. In: *Wireless Communications and Networking Conference (WCNC), 2016 IEEE*, 1–6. IEEE.

Goggin, G. (2012). *Cell Phone Culture: Mobile Technology in Everyday Life*. Routledge.

Jordan, E. K. (2017). *U.S. Patent No. 9,665,476*. Washington, DC: U.S. Patent and Trademark Office.

Kaleem, Z., Rehmani, M.H., Ahmed, E. et al. (2018). Amateur drone surveillance: applications, architectures, enabling technologies, and public safety issues: Part 2. *IEEE Communications Magazine* 56 (4): 66–67.

Lu, N., Cheng, N., Zhang, N. et al. (2016). Wi-Fi hotspot at a signalized intersection: cost-effectiveness for vehicular internet access. *IEEE Transactions on Vehicular Technology* 65 (5): 3506–3518.

Mariappan, P.M., Raghavan, D.R., Aleem, S.H.A., and Zobaa, A.F. (2016). Effects of electromagnetic interference on the functional usage of medical equipment by 2G/3G/4G cellular phones: a review. *Journal of Advanced Research* 7 (5): 727–738.

Merwaday, A. and Guvenc, I. (2015). UAV assisted heterogeneous networks for public safety communications. In: *Wireless Communications and Networking Conference Workshops (WCNCW), 2015 IEEE*, 329–334. IEEE.

Rao, S.K. and Prasad, R. (2018). Impact of 5G technologies on smart city implementation. *Wireless Personal Communications* 100 (1): 161–176.

Samba, A. S. (2014). *U.S. Patent No. 8,868,725*. Washington, DC: US Patent and Trademark Office.

Schiller, E., Kalogeiton, E., Braun, T. et al. (2017). ICN/DTN for public safety in mobile networks. In: *Wireless Public Safety Networks*, vol. 3, 231–247.

Schneider, P. and Horn, G. (2015, August). Towards 5G security. In: *Trustcom/BigDataSE/ISPA, 2015 IEEE*, vol. 1, 1165–1170. IEEE.

Tiwari, S., Lane, M., & Alam, K. (2015). The challenges and opportunities of delivering wireless high-speed broadband services in Rural and Remote Australia: A Case Study of Western Downs Region (WDR). Australian Conference on Information Systems, https://arxiv.org/ftp/arxiv/papers/1606/1606.03513.pdf.

Tsai, W.C., Huang, N.T., Lin, S.H., and Chiang, M.L. (2017). An implementation of relay function based on LTE technology for multi-UAV communication. In: *International Conference on Smart Vehicular Technology, Transportation, Communication, and Applications*, 365–373. Cham: Springer.

10

Fifth Generation (5G) Cellular Technology

10.1   Introduction

Cellular wireless communication is normally categorized in generations that are normally denoted by G. The generation refers to metamorphism in the primary nature of the services offered, frequency bands and the non-backward compatibility of the individual transmission technology. Since the establishment of the first generation (1G), we can say that movement from one generation to another takes approximately 10 years. The first generation was unleashed in 1981, which lead to the provision of 1G and later 2G, which were analog networks. After that, the third generation (3G), which had support for multimedia and a spread spectrum, was unveiled. Drawbacks in the third generation led to the release of the fourth generation (4G), which was introduced in 2011 and supported all IP-switched networks.

The world has witnessed growth in mobile technology thus requiring effective innovations in both information and communication technologies to manage the ever-growing demand for wireless devices. The high demand for mobile data services through wireless internet and smart machines have exceeded the capability of the current 4G networks thus triggering research into the next generation of terrestrial mobile telecommunications known as 5G. The 5G network is to be an improvement on the current 4G, which is seen as a multitude of heterogeneous systems that are programmed to interact through a horizontal IP-centric architecture. The 5G architecture is to be built on a reconfigurable multi-technology core. The architecture is anticipated to be a combination of several emergent technologies such as cloud computing, nanotechnology, cognitive radio, and other technologies based on an all-IP platform. However, incorporating these new technologies is likely to cause challenges in the development of the 5G network.

The development of the fifth generation of wireless systems is underway and is expected to be released beyond 2020. The purpose of this chapter is, therefore, to analyze the development of the fifth generation of wireless communication systems and the network of tomorrow. This will be accomplished through discussing the proposed 5G network architecture, wireless technologies used in the development of the network generation, lessons from the previous generations and other important features in the development of the fifth generation of wireless communication.

Public Safety Networks from LTE to 5G, First Edition. Abdulrahman Yarali.
© 2020 John Wiley & Sons Ltd. Published 2020 by John Wiley & Sons Ltd.

10.2 Background Information on Cellular Network Generations

10.2.1 Evolution of Mobile Technologies

Evolution can be defined as the essence of the effect that is left by every being or technology. The computer network has greatly evolved since the invention of telegram to the network of the future and 5G communication. Mobile technologies are, therefore, categorized into generations that are 10 years apart. The fifth generation is an improvement on the previous generations. Drawbacks of the latter generations can be used to develop a better fifth generation.

10.2.1.1 First Generation (1G)

This generation of the cellular network was invented in the 1980s and was developed behind the idea that geographical areas are divided into small locations known as cells (10–25 km). Every cell served as a "base station" where frequency reuse was used in the nearby cells. Every device supported by the 1G network was analog and usually used what was popularly known as cellular phone technology that was working in the frequency band of 150 MHz. 1G networks comprised of a number of mobile technologies such as advance mobile telephone systems (AMTS), improved mobile telephone service (IMTS), and mobile telephone systems (MTS).

1G networks had several problems that limited their use and prompted advancement to a more advanced technology that reduced these drawbacks. Analog cellular phones were very insecure in that anyone with an all-band radio receiver that was connected to a computer could record the 32-bit serial numbers and the subscriber's phone number and listen to the communication. This loophole was used in many cases. Additionally, there was stealing of airtime and reprogramming of stolen phones so that they could be resold.

10.2.1.2 Second Generation (2G) Mobile Network

The second generation telecommunication networks were launched first in Finland in 1991 on the Global System for Mobile Communications (GSM) standards. 2G was more advanced as it used digital signals for voice transmission that had a speed of 64 kbps. Additionally, this mobile technology supported a short message service (SMS) that used a bandwidth range of 30–200 KHz. The 2G network comprised of a number of advanced mobile technologies such as GPRS, enhanced data rates for global system for mobile communications evolution (EDGE), code division multiple access (CDMA), and GSM. These technologies made this generation better than its predecessor in that it supported digital encryption and had higher penetration, making it a more efficient network spectrum. This mobile generation is still used in many parts of the world today.

This network generation had a number of issues that led to its advancement. For instance, the network system used digital signals that had a rough decay curve as opposed to analog signals thus making the signal weaker in areas that were less populated.

10.2.1.3 Third Generation (3G) Mobile Network

This generation of telecommunication systems uses a wide brand wireless network whose clarity is greatly increased. The voice calls are normally interpreted using circuit switching while data are transmitted through packet switching technology. This generation has increased capabilities such as television/video access and global roaming. 3G networks operate at 2100 MHz and have a bandwidth of 15–20 MHz, thus offering high-speed access to internet services such as video streaming and video chatting. This generation has seen the world compressed into a digital village where one can communicate to a person from any part of the world. 3G is comprised of Bluetooth, wide-band CDMA, universal mobile telecommunication systems (UMTS), wireless local area network (WLAN), and high-speed downlink packet access (HSDPA). 3G also supports additional functionalities such as a multi-media messaging service (MMS), video conferencing, 3D gaming, video calls, and multiplayer gaming.

The high adoption rate of 3G has made its price come down, although it is still very expensive as compared to 2G technology. Additionally, the power consumption of 3G technology is high due to high bandwidth transmission, thus leading to a reduced battery life of 3G devices. The high bandwidth transmission has consequently led to big network overload and therefore is disadvantageous to customers.

10.2.1.4 Fourth Generation (4G) Mobile Network

The 4G communication system is not meant to provide voice calls and other 3G capabilities only but is also meant to provide broadband network access to mobile devices. Using 4G, devices can access applications such as HD mobile television, cloud computing, IP telephony, and video conferencing.

The 4G mobile network comprises of the Long Term Evolution (LTE) standard that is based on GSM/EDGE and UMTS/high-speed packet access (HSPA), the Third Generation Partnership Project (3GPP), orthogonal frequency digital multiplexing (OFDM), multiple input, multiple output (MIMO), WiMAX, and mobile broadband wireless access (MBWA).

Nevertheless, this mobile network has a number of drawbacks that limit its usability. For instance, the issue of different frequency bands limiting the use of similar devices in different continents. Another serious issue with 4G is on how to make higher bit rates available in a large portion of a cell. This problem is being solved using a macro-diversity technique or beam-division multiple access (BDMA). Table 10.1 summarizes some characteristics of theses mobile generations.

10.2.1.5 Fifth Generation (5G)

This mobile technology is based on LTE, and it is in its early stages of development with an aim to revolutionize mobile communication by coming up with systems that adequately eliminate current problems that are being experienced with the existing mobile systems. The following are other factors that 5G is aiming to implement better than its predecessor other than throughput:

- Reduced battery consumption
- Increased range of coverage and higher data rate throughout the cell
- A high number of concurrent data transmission paths and hand over
- Interactive multimedia services such as voice call, video call, the internet, and other broadband services that are more effective for bidirectional traffic statistics
- Mobility data rate averaged at over 1 Gbps and a higher broadcast capacity of over 65 000 connections at a time

Table 10.1 Summary of comparison between the five mobile communication generations.

Technology	1G	2G	3G	4G	5G
Deployment	1970–1980	1990–2004	2004–2010	Now	Soon (probably 2020)
Throughput	2 kbps	64 kbps	2 Mbps	1 Gbps	Higher than 1 Gbps
Technology	Analog cellular technology	Digital cellular technology	CDMA 2000 (1xRTT, EVDO) UMTS, EDGE	Wi-Max LTE Wi-Fi	WWWW (coming soon)
Service	Mobile telephony (voice)	Digital voice, SMS, higher capacity packetized data	Integrated high-quality audio, video, and data	Dynamic information access, wearable devices	Dynamic information access, wearable devices with AI capabilities
Multiplexing	FDMA	TDMA, CDMA	CDMA	CDMA	CDMA
Switching	Circuit	Circuit, packet	Packet	All packets	All packets
Core network	PSTN	PSTN	Packets N/W	Internet	Internet

FDMA, frequency division multiple access; TDMA, time division multiple access; PSTN, public switched telephone network.

- Improved security features such as better cognitive software development radio (SDR)
- Worldwide wireless web (WWWW) based applications that have full multimedia capability
- Artificial intelligence aided applications
- High resolution for cellphone users and bi-directional large bandwidth shaping.

10.3 Fifth Generation (5G) and the Network of Tomorrow

5G is the proposed next phase of mobile telecommunication standards after the fourth generation. The proposed network aims to improve internet connection speed and other improvements on the current 4G network. History has it that it takes 10 years to move from one mobile generation to another. Therefore, it is proposed that the 5G network is likely to be unveiled in the early 2020s.

The fifth generation (5G) wireless networks will be designed to provide higher data rates, handle more traffic in a dense environment, reduce end-to-end latency, support devices with ultra-low energy consumption, reduce total deployment cost, and provide enhanced overall customer quality-of-experience (QoE). Some of these requirements may be in conflict hence the designer of the network of tomorrow must exercise trade-offs to achieve an architecture that supports a specific application. This requires the 5G network for tomorrow to be cognitive and cooperative. Cognitive means that all the nodes need to be aware of their environment and possibly other node environments. Cooperative means that these nodes are able to work closely with surrounding nodes to improve system performance and enhance user experience.

Nevertheless, 5G will not happen in the manner and timescale that we have been told. 5G will only justify itself if it enables new behaviors and business cases in order to bring sufficient revenue and sustain development effort. Harmonization of the 5G spectrum worldwide is also a key, otherwise it risks losing out in the economies of scale that 5G handsets, cars, and the Internet of Things (IoT) need. Recently, the FCC took steps to facilitate mobile broadband and next-generation wireless technologies by proposing to allocate a spectrum for 5G above 24 GHz. The impact of such a proposal on the harmonization of the spectrum and hence the deployment schedule of 5G remains to be seen. Some chip vendors claim that 5G is almost here. The reality is some 5G features that are compatible with LTE and LTE-Advanced will be deployed first. In most cases, 5G components will sit happily inside a sub-6 GHz LTE and LTE-A networks. In other cases, 5G will become an extension of the communication platform created by the LTE common air interface. In short, 5G is about many things and will be deployed in phases depending on which feature is the least expensive and the most impactful for operators to implement first.

Several techniques, which will enable and accelerate the deployment of 5G mobile communications networks for tomorrow are:

- Cooperative multifunction communication where a wireless communication node will include a controller configured to determine capabilities of at least one remote access terminal, and delegate at least one task to the at least one remote access terminal based on the capabilities.
- The proxy concept in wireless networks where one node will agree to monitor the network on behalf of another node that is in power conservation mode. The proxy node will wake up the intended node and relay the directed message through a low power communication link using out-of-band radio.
- Cooperation between multiple radios in wireless devices.
- The concept of assisted network acquisition and system determination. Rather than scanning one or more frequency bands to discover local wireless network information, a first device may send a request for local wireless network information to a nearby second device. Such request for network information may be a specific request for information about a particular type of wireless network (e.g. 5G, LTE, WLAN), a specific request for one or more networks associated with a particular service provider.
- The concept of network energy savings. 5G wireless networks will be characterized by diverse bands and bandwidths that encompass licensed, unlicensed, and authorized shared spectrum resources.
- The concept of data mules that provide or enable a dynamic cooperative wireless data delivery service based on dynamic proximate locations of mobile nodes in wireless networks.

10.3.1   5G Network Architecture

To effectively address the challenges being experienced by the previous cellular network generations, we need a dramatic overhaul in the design of cellular architecture. Research shows that wireless device users stay indoors most of the time, precisely 80% of their time, while they spend 20% of their time outdoors and in motion (Chandrasekhar et al. 2008). The current cellular architecture is designed to use an outdoor BS positioned in the middle of a cell communicating with mobile users, regardless of whether they are indoors or outdoors. This means that for indoor users communicating using the outdoor BS, signals have to pass through walls, which causes high signal loss, consequently leading to reduced data rates, energy efficiency, and spectral efficiency of the entire wireless transmission.

The primary idea in developing 5G cellular architecture is the separation of indoor and outdoor cases so as to minimize penetration loss. Use of the two scenarios will be assisted by the use of massive MIMO and distributed antenna system (DAS) technology (Rusek et al. 2013). A geographically distributed antenna array made of hundreds of antenna components will be deployed in the implementation of DAS. The goal of many MIMO systems is to make use of the large capacity gains that are likely to arise after deployment of large arrays of antennas. The existing outdoor BS systems will be fitted with a large antenna array that has antenna components placed around the cell and connected to the BS using optical fibers. This will enable the BS to benefit from both massive MIMO and DAS technologies. Outdoor users are usually equipped with a lower number of antennas, but they can collaborate with others to form virtual large antenna array that will be connected with a BS antenna array to come up with massive virtual MIMO links (Damnjanovic et al. 2011). This architecture will also involve installing large antenna arrays outside every building for streamlined communication with outdoor BSs via line of sight (LOS) elements.

In this architecture, the indoor mobile users will only have to communicate with indoor wireless access points and the large antenna arrays will be mounted outside the building will communicate with the BS. Many technologies can be used to handle these short-range communications with massive data rates such as Wi-Fi, ultra-wideband (UWB), femtocell and visible light communications (VLC).

The 5G cellular architecture design needs to be a heterogeneous one fitted with microcells, macrocells, relays, and small cells. Additionally, the design has incorporated the mobile femtocell (M-Femtocell) concept to support high mobility users such as those in high-speed trains and vehicles. The M-Femtocell concept is designed to combine both concepts of femtocell and mobile relay for high output. This concept uses M-Femtocells that are located within the vehicle to communicate with the users inside the vehicle and then communicate with large antenna arrays that are located outside the vehicle to relay the communication to the outdoor BSs (Saquib et al. 2012). In this scenario, the M-Femtocell and its users are viewed as one unit by the outdoor BS, while from the user point of view, the M-Femtocell is viewed as a regular BS. Using this concept, users within the vehicle can also experience high data rate with a decreased signaling overhead. The diagram in Figure 10.1, is a representation of the proposed 5G heterogeneous cellular architecture.

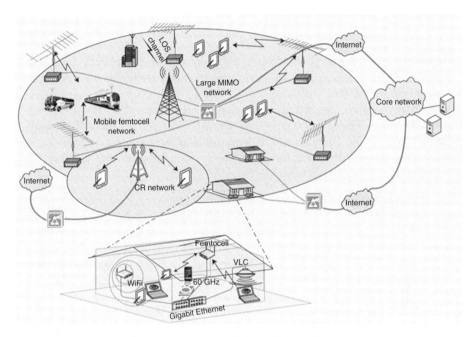

Figure 10.1 The proposed 5G heterogeneous wireless cellular architecture.

10.3.2 Wireless Communication Technologies for 5G

Based on the proposed architecture above, I am going to discuss some of the promising wireless technologies that are designed to allow the 5G network to fulfill its performance requirements. The purpose of these technologies is to allow increased data rates while increasing efficiency and reducing implementation cost. According to the Shannon theory, we can calculate the approximate total system capacity using the following equation:

$$C_{\text{Sum}} = \sum_{\text{HetNets}} \sum_{\text{Channels}} B_i \log_2 \left(1 + \frac{P_i}{N_p}\right)$$

where B_i is the bandwidth of the ith channel, N_p the noise power and P_i the signal power of the ith channel. From the given equation, we see that the total system capacity of the network is equivalent to the sum of all heterogeneous network and sub-channel capacity. This is a clear indication that if we wish to increase the capacity of the network, we need to increase the number of sub-channels (via MIMO) and network coverage (by increasing heterogeneous networks using microcells, macrocells, and M-Femtocells). The following are some of the key technologies implemented in 5G cellular network architecture.

10.3.2.1 Massive MIMO

MIMO technology depends on multiple antennas to simultaneously transmit multiple streams of data in a wireless communication system (Gesbert et al. 2003). MIMO systems are designed to have multiple antennas at both the transmitter and the receiver. These antennas can either be co-located or distributed in various applications. Sometime MIMO can be used to communicate with several terminals simultaneously. In this case, it is known as multi-user MIMO, MU-MIMO. MU-MIMO improves a cellular system in the following four dimensions:

- Increase data rates. This happens because the many antennas make data streams sent out more independent making it easier to serve the many terminals simultaneously.
- Enhanced reliability. The many antennas create many distinct paths that the radio signals are propagated over increasing the system reliability.
- Improved energy efficiency. This is because a base station (BS) can focus all the energy emitted into a specific direction where the terminal is located.
- Reduced interference. The base station purposely avoids transmitting into directions where spreading interference would be harmful.

All improvements cannot be implemented simultaneously because there are specific requirements for propagation conditions that need to be met to achieve the above bene-fits. Implementation of MU-MIMO technology for wireless communication is in the last phase of development and already been incorporated in the latest wireless broadband standards such as LTE-A and 4G LTE (Dahlman et al. 2013). The bigger the number of antennas that a terminal has, the better the performance. The LTE-A, one of the recent technological advancement, allows up to eight antennae.

Massive MIMO is an advanced version of MIMO that aims at increasing the mag-nitude of the current version of MIMO. Massive MIMO is a system that uses antenna arrays that have over a hundred dedicated antennas that are designed to serve tens of terminals in the same time-frequency resource. The primary goal of massive MIMO is to reap all the benefits of conventional MIMO while taking it to a higher productivity level. It is meant to enable the development of future mobile broadband networks that are energy-efficient, robust, secure, and will use the spectrum efficiently. It is, there-fore, incorporated in the 5G network architecture to enable connection of IoT, cloud computing, and other network infrastructure.

Massive MIMO depends on spatial multiplexing that also depends on the ability of the base station to have enough channel knowledge during both downlink and uplink. It is easy to achieve this during the uplink by having the terminals send pilots that are based on the base station estimates and channel response to every terminal. This is difficult to implement in the downlink because pilots have to be mutually orthogonal between the antennas, or the number of channel response estimated by every terminal must be pro-portional to the number of base station antennas. The solution would be to operate the system in time division duplexing (TDD) mode while relying on the reciprocity between the downlink and the uplink channels.

Massive MIMO relies on phase-coherent but a very simple computational method of processing signals from every antenna at the base station. Therefore, this technology has realized many benefits over conventional MIMO. The following are some of the benefits:

- Increased capacity of about 10 times while simultaneously improving the radiating energy efficiency 100 times.
- Increased robustness to both intentional jamming and man-made interference. Jam-ming wireless system is a common cyber-security issue that has a great impact on wireless communication channels. Due to the many channels available in massive MIMO, there is a great sense of freedom that is used to cancel signals from intentional jammers.

- Massive MIMI simplifies the multiple-access layer. Due to the large numbers, the channel hardens to a point such that frequency-domain scheduling is ineffective. Therefore, each subcarrier in the massive MIMO system will have the same channel gain. Each terminal can, therefore, be given a whole bandwidth making most of the physical-layer control signaling redundant.
- Massive MIMO is built with inexpensive, non-power consuming components. In massive MIMO design, hundreds of low-cost amplifiers that have a milli-Watt range of output power are used in the place of ultra-linear 50 W amplifiers that are used in conventional MIMO systems.
- Wireless communication involves the transmission of signals through the air. As the signals travel through the air, their energy is greatly reduced by fading. This happens when the signal transmitted from the base station is propagated through multiple paths and reaches the terminal resulting in destructive interference. The fading of the signal makes it hard to build low-latency wireless links. However, massive MIMO can avoid fading dips by the large number of signals and beam-forming. Massive MIMO, therefore, significantly reduces latency on the air interface.

10.3.2.2 Spatial Modulation

Spatial modulation is a novel MIMO technique that has been proposed for low-complexity implementation of MIMO systems without reducing the system performance (Chin et al. 2014). Spatial modulation normally encodes part of the data that is to be transmitted to the spatial position of each transmit antenna array instead of transmitting the multiple data simultaneously. The antenna array is, therefore, used as the second constellation diagram, which is normally used to boost the data rate for a single-antenna wireless system. In spatial modulation, only one antenna is active during transmission and others are idle. Spatial modulation is a combination of amplitude modulation and space shift keying modulation.

Spatial modulation is a technology that is introduced to solve three of the weaknesses in the conventional MIMO systems, i.e. multiple RF chains, inter-antenna synchronization, and inter-channel interference. Spatial modulation systems can also be designed with low-complexity receivers that are configurable to accept any number of receive and transmit antennas. Many of the existing spatial modulation systems are a single receiver. Therefore, research into a multi-user spatial modulation is geared toward the realization of 5G wireless communication systems.

10.3.2.3 Machine to Machine Communication (M2M)

As the world grows smarter, the introduction of autonomous machines in these smart networks is inevitable. 5G technology, therefore, will involve communication between machines in the smart network. Such communication is referred to as "machine to machine" communication. This technology is, therefore, going to be phenomenal in the telecommunication world as it will enable the connection of cellular devices.

However, there are several challenges associated with M2M communication that involves the operation, power consumption, and complex requirements. For instance, it is hard to handle the flow of massive data that are generated and transmitted by these machines. This, therefore, requires complex network architectures that are easy to extend and modify. M2M communication together with the IoT, therefore, are pillars of an intelligence core that is an important component in enabling seamless connectivity in a 5G context (Chih-Lin et al. 2014). Despite the benefit of M2M technology, there is a need for an intelligent core that can be able to handle the big data that is collected and transmitted through this technology.

10.3.2.4 Visible Light Communication (VLC)

This type of communication uses off-the-shelf white light emitting diodes as signal transmitters and avalanche photodiodes as the signal receivers. VLC is normally used to enable systems that emit light to have the ability to provide broadband wireless data connectivity. This means that the information is carried by light intensity. Therefore, the signal that is relaying information has to be strictly positive and real-valued. Research has shown that a single LED can achieve a data rate of 3.5 Gbps. Additionally, VLC systems are not affected by fast fading effect because the wavelength used is significantly small as compared to the detector area.

10.3.2.5 Green Communications

The 5G wireless communication systems are proposed to be greener, which is achieved by minimizing energy consumption. Due to the environmental degradation activities taking place around the world, manufacturers are doing their best to ensure they come up with environmentally friendly products and wireless communication operators are not an exception (Chih-Lin et al. 2014). One way that development of 5G cellular network developers is working at is to reduce the power consumption of the systems, which will consequently contribute to the reduction of carbon dioxide emissions to the environment. This can be achieved by capitalizing in indoor communication systems that are energy efficient. Additionally, separation of indoor systems from outdoor systems means the BS will not have a problem in allocating radio resources and therefore can transmit with less power, leading to a significant reduction in power consumption.

10.3.3 5G System Environment

In order to provide the high-speed services required by the 5G architecture, manufactured systems will need to be context aware by making use of context information in a real-time manner depending on the applications, devices, network, and the user environment. Context awareness will allow improvements in the efficiency of existing services and use them to provide more personalized and user-centric services. For instance, mobile networks will need to be conversant with system application requirements, and specific ways to adapt its application flow processes so as to meet the user needs. This can be achieved by introducing new interfaces between application and network layers so as to deliver the best capability to the users. The following are some of the areas that aware context adaptation must follow:

- Device-level context information on CPU load, battery state, and specific device features.
- Application-level context information such as web browsing, cloud-based application interactivity, video capability, and system support for major applications.
- User-level context information which includes user activity, user preference on quality, user location, and level of distraction.
- Environmental context information such as lighting, motion, and close devices.

- Network context information such as load, backhaul quality, alternative paths, and network throughput.

Unlike in the previous generations, 5G cellular technology needs efficient ways to generate and share context information between devices, application, and network for optimal performance.

10.3.4 Devices Used in 5G Technology

Before network infrastructure improvements to accommodate the transition from 4G to 5G is device transition to ensure they accommodate the user and the network requirements. The current devices are supposed to evolve in size, function, and form to meet the 5G technological requirements. However, it is important to note that 5G deployment will not be possible at the same time in all areas. Therefore, 5G devices should be downward compatible so as to use lower network infrastructures. Thus, a device that supports 5G RAT should at least be able to support LTE, wideband code division multiple access (WCDMA), HSPA, Wi-Fi, and Bluetooth.

The demand for radio access technology (RAT) powered devices has greatly increased with wearables and machine-type devices such as smartphones and tablets entering the market. This has caused much pressure on device chipset and platform suppliers. This progressive rise in form factor diversity has led to the production of different platform support that in turn makes transceiver complexity a real challenge in 5G device design.

The 5G devices will need to raise the number of supported antenna ports from two to four so as to be able to support higher spatial multiplexing and increase cell edge performance. This has already been implemented in LTE and other 5G devices that use RAT architecture. Implementation of these changes is coupled with reduced device volume and increased power need. However, the slow progress in battery chemistry is derailing the implementation of wideband antennas that can be shared between multiple RATs.

The devices will also be able to increase inference suppression through advanced signal processing that is done in massive MIMO technology and MU-MIMO. 5G RAT devices will also operate in millimeter-wave frequencies that perfectly handles problems in the antenna array and high-speed transceiver design. The primary challenge with millimeter-wave integration its high cost and low power operation. This means that millimeter-wave in 5G RAT architecture will only be able to operate only when the resulting energy consumption per bit is optimal, which can only be enabled by the use of sophisticated connection managers.

5G technology devices, Figure 10.2, will support a large array of device location capabilities that include high processing, ephemeris assistance data, GPS, and global navigation satellite system (GLONASS) satellites. The devices will also require supporting new frequency bands, such as GPS L2 and L5 bands, and pilot signals that require 3G-like channel and time-delay estimation. Additionally, 5G devices are expected to support Bluetooth-based ranging and RF techniques that will be embedded in a holistic location engine that will incorporate technologies such as crowdsourcing and geo-fencing.

10.3.5 Market Standardization and Adoption of 5G Technology

Over the past few decades, mobile technology has experienced rapid development in terms of speed and efficiency that can be viewed in a service-driven perspective. The

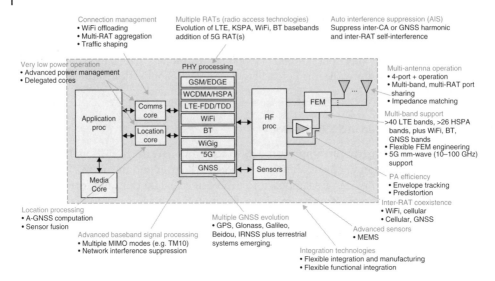

Figure 10.2 Device technologies for 5G network.

commercial value of every technology depends on the services and applications required by users. Due to the large number of individual parties involved in the development process of the 5G network, there is a need to come up with standardizing bodies that will ensure device and technology compatibility is adhered to in the development process (Di Renzo et al. 2014). There is, therefore, co-evolution between the market and technology advancement. This can be observed in the action–response that is visible in the progressive sequence of supply and demand.

The service-driven perspective is the one that makes a regulator focus more on the non-technical issues while the technical matters are left to the technology implementer. For instance, in the case where the market expects advanced apps to support their financial transactions, the regulator, in that case, will not be concerned with the technology behind the applications. In this the regulator lets the operator come up with a suitable technology to meet the market demand. The regulator, in this case, will only manage non-technical matters such as coordinating with financial institutions.

In 3G and 4G technologies, it is clear that all applications were targeted at enriching communications between people. This means that regulators' policies were entirely targeted at improving personal communications. Some of the policies that have been set are to ensure market fairness and subscriber privacy protection. 5G technology, however, is not entirely targeting the human market. This is because by the time 5G is rolled out, there will be excessive growth in IoT and M2M applications and personal communication will already have reached a saturation level.

This, therefore, means that policymakers in the development of 5G technology should put into consideration the effect that will be created by the interconnected machines and the IoT. The IoT is a technology where one will be able to communicate with machines such as a car, watch, utensils, air conditioner, and other devices on a daily basis. 5G vision is, therefore, designed to connect devices to the human network so as to make them manageable and attain maximum utilization. Due to the multiple connections available between various devices, the issue of privacy and security has become an important issue to be put into consideration. The introduction of market standardization is, therefore, a serious problem after the introduction of the IoT platform. This is because such market changes are complex and will require the policymakers to consider the broader dimension of social and cultural factors.

10.3.6 Security Standardization of Cloud Applications

Development of 5G technology will trigger profound cloud-based applications and services that emerge at unimaginable levels. Cloud computing provides many advantages to businesses such as flexible cost structure, efficiency, and scalability (Osseiran et al. 2014). The promising advantages of cloud computing have lured many organizations, small and large, to venture into it. However, cloud technology can be viewed as a disruptive technology because it is likely to change greatly the traditional way of availing computing services to organizations and individuals (Zhang et al. 2015).

Data security in cloud computing is an important aspect that need to be put into consideration and standardized. There are many aspects of information security that need to be harmonized such as network security, data security, and perimeter security for cloud applications to be entirely secure. At the global layer, the standardization reference is listed in ITU-T Recommendation X.805 "Security architecture for systems providing end-to-end communications, Data Security Framework Rev1.0" issued by the Open Data Center Alliance (ODCA 2014). The International Telecommunication Union (ITU) states that there is a standardization gap between developing and developed countries when it comes to implementation.

Figure 10.3 shows a block diagram of data security recommendations. The primary principle of the above framework is to provide security for the whole application by considering threats and vulnerabilities. This system security system is based on a layering and plane concept that aims at obtaining security end-to-end on every layer. As depicted in the above diagram, security aspects are divided into access control, authentication, non-repudiation, data confidentiality, communication security, data integrity, availability, and privacy. However, it is hard for regulators to ensure security at every layer.

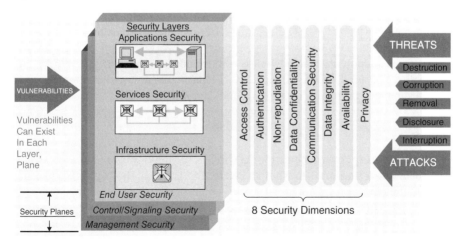

Figure 10.3 Security architecture X.805 ITU-T (Zachary, 2003). Source: ITU-T (2003).

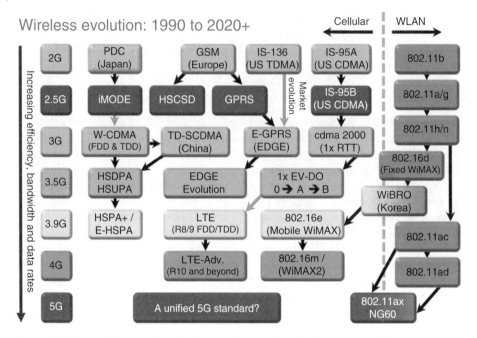

Figure 10.4 The cellular tree of standardization. Source: Keysight Technologies -5G.

10.3.7 The Global ICT Standardization Forum for India (GISFI)

GISFI is an Indian standardization body that was established in 2009 to set standards in information and communication technologies (ICTs) and related areas of application such as energy, wireless robotics, telemedicine, and biotechnology. The purpose of this forum is to create coherence between the world standardization processes and ICT achievements in India (Prasad 2013). It is a fundamental step in growing the tree of standards that is aimed at reaching 5G. Figure 10.4 shows a roadmap of wireless mobile communications standards for various coverage and speed range. The Global ICT Standardization Forum for India (GISFI) uses a technology concept of a system that is commonly known as the wireless innovation system for dynamically operating mega-communications (WISDOM).

GISFI presents a vision of a future wireless communication system that would be universally deployed as a pillar for enabling a smart tele-infrastructure that offers services and applications of more than a Tera/bit/second data rate.

10.3.8 Energy Efficiency Enhancements

One of the engineering requirements for 5G wireless technology is to increase the data rate while decreasing the power consumption rate. This is a great motivation in ensuring this new technology conforms to green world requirements. To achieve this, different mechanisms have been put in place to ensure production of low power consumption devices that can run all 5G application and services. The proposed decrease in power consumption should be equal to the per-link data rates which are being postulated to be 100X. Energy consumption of a given consumption chain is normally given in bits/Joule or Joules per bit (Chin et al. 2014). Therefore, to ensure limited energy consumption of 5G systems, the following areas have to be considered:

- Resource allocation. High energy savings can be attained by adjusting the devices to accept reduced data rates. This has already been effected in HetNets, which has introduced energy-efficient coordinated beam-forming.
- Network planning. Energy saving network design is another aspect that can be used to reduce energy consumption in 5G systems. One technique that is used in the described network architecture is minimizing the number of BSs for a given coverage target or by using an adaptive BS algorithm that hibernates parts of the network that are idle.
- Use of renewable energy. Using solar-powered BSs is another method that has been proposed as a way of reducing the energy requirements of the system. This mechanism is very useful in developing countries that do have a reliable power grid.
- Hardware solutions. Most power issues can be solved through hardware optimization. One of the ways that telecommunication engineers have proposed for 5G devices is to use low-loss antennas, and antenna muting.

10.3.9 Virtualization in the 5G Cellular Network

To realize effective development in wireless traffic and services, there is a need for virtualization, which has already been successfully implemented in a wired network (such as VPN) and wireless network. With the use of network function virtualization (NFV) the proposed 5G wireless network infrastructure can be decoupled from the services it provides so that similar applications can be made to use a personalized infrastructure, thus maximizing their utilization. NFV, therefore, provides a foundation for the development of design principles that can be used to come up with software-defined 5G wireless networks. Wireless network virtualization can also be used to provide an easier transition between emerging and legacy technologies by isolating each from the major network. Research into incorporating virtual network in cellular wireless communication can be traced back to eNodeB that was incorporated in 3GPP as part of LTE (Felser and Sauter 2004).

In virtualization of a wireless network, virtual resources are formed by cutting physical resources into a large number of virtual slices. A single slice must have every virtual entity derived from each element in the wireless network infrastructure (Demestichas et al. 2013). The most common way of virtualizing a wireless network is the use of an information-centric wireless virtualization controller. This process involves different steps, i.e. abstraction, physical resource allocation, virtualization, slicing, isolation, and assignment (Demestichas et al. 2013).

10.3.10 Key Issues in the Development Process

To cope with the development issues affecting the 5G cellular network, we have to look at challenges facing every major technology used in the development process. As described above, various new technologies are being proposed and developed to realize 5G networks. Before looking at the drawbacks posed by individual technologies, it is important to state that the integration of all the technologies is likely to cause a serious problem in the development process. Regardless of the drawbacks experienced during the development stages, it is worth stating that 5G cellular network will provide thrill for the whole mobile technology. The new technologies introduced are meant to reduce the drawbacks experienced with the previous generations such as reliability, robustness, throughput, and power consumption (Chin et al. 2014). The following are some of the development issues likely to be experienced during the 5G network development process.

10.3.10.1 Challenges of Heterogeneous Networks

- Inter-cell interference. This problem is encountered when unplanned small cells are deployed in areas where network operators have no control. Additionally, there is likely to be distortion of signals where small cells are combined with traditional macro cells leading to inter-tier inference. This type of inference can be avoided by the use of advanced power control coupled with planned resource allocation.
- Device discovery problem. In D2D communication, there is a possibility of low link up where there are many devices. When these devices are operating at the frequency, it can be problematic to set it up and maintain a link with multiple devices.
- Distributed interference coordination. In case there is little or no coordination between different WLANs, a mechanism to avoid distributed need be devised when installing a new access point. This will be important because more devices around those areas will be seeking to connect in their endeavor to complement their bandwidth.

10.3.10.2 Challenges Caused by Massive MIMO Technology

As stated earlier, Massive MIMO is a technology that is being devised to use a large array of antennas as compared to those being used today. Though this technology is very useful at high frequencies and where antenna components can be miniaturized, it has a number of challenges:

- The challenge of pilot contamination. The fact that MIMO suffers from pilot contamination caused by other cells means that some critical adjustment will need to be done for MIMO to deliver full performance.
- Massive MIMO will require fast processing algorithms to sort and process the massive data arriving from the many RF chains.

10.3.10.3 Big Data Problem

Like in other marketing industries, big data will be a source of many challenges to 5G wireless technology. The main aim of 5G technology is to provide high-speed data capability which in turn requires complex and efficient infrastructure to maintain this data deluge. The introduction of the IoT and M2M application in 5G development will mean an explosion of such data. This is, therefore, a serious technical challenge to the developers. Additionally, there is a likelihood of a new network architecture emerging to aid big data handling.

10.3.10.4 Shared Spectrum

Cognitive radio has been thought of as a solution to the issue of spectrum shortage for some time, but there are also concerns about the effect it has on license holder of the spectrum and the primary user. This challenge can be solved by introducing the authorized spectrum access (ASA) technique that is used to allow only authorized users based on specific rules set by the holder of the spectrum license. This technique is effective in ensuring the under-utilized spectrum are utilized to the maximum thus improving signal quality to the primary user.

10.4 Conclusion

5G technology is a brilliant technology where connectivity and networking innovations are used to increase the personalization of cellular devices. By use of technologies such as M2M technologies, networking technologies, data mining, cognitive radio, decision-making technologies, cloud computing, advanced sensing technologies, a more powerful and flexible network, 5G, has been realized. To ensure swift adoption, GISFI has taken the mandate for 5G standardization by defining recommendations for the realization of the WISDOM concept (Andrews et al. 2014).

In this chapter, I have provided a comprehensive overview of the different new technologies that have been set as the pillars to the realization of a 5G network and network of tomorrow. Additionally, I have analyzed market standardizations that have been proposed for 5G cellular technology. In spite of the many exemplary technologies that have been proposed and put to a test to ensure standardization and effectiveness of the proposed innovation, 5G, there are several challenges that developers have to deal with before realizing the visionary output.

References

Andrews, J.G., Buzzi, S., Choi, W. et al. (2014). What will 5G be? *IEEE Journal on Selected Areas in Communications* 36 (2): 1065–1082.

Chandrasekhar, V., Andrews, J.G., and Gatherer, A. (2008). Femtocell networks: a survey. *IEEE Communications Magazine* 46 (9): 59–67.

Chih-Lin, I., Rowell, C., Han, S. et al. (2014). Toward green and soft: a 5G perspective. *IEEE Communications Magazine* 52 (2): 66–73.

Chin, W.H., Fan, Z., and Haines, R.J. (2014). Emerging technologies and research challenges for 5G wireless networks. *IEEE Wireless Communications* 12 (2): 106–112.

Dahlman, E., Skold, J., and Parkvall, S. (2013). *4G: LTE/LTE-Advanced for Mobile Broadband*. Academic Press.

Damnjanovic, A., Montojo, J., Wei, Y. et al. (2011). A survey on 3GPP heterogeneous networks. *IEEE Wireless Communications* 18 (3): 10–21.

Demestichas, P., Georgakopoulos, A., Karvounas, D. et al. (2013). 5G on the horizon: key challenges for the radio-access network. *IEEE Vehicular Technology Magazine* 8 (3): 47–53.

Di Renzo, M., Haas, H., Ghrayeb, A. et al. (2014). Spatial modulation for generalized MIMO: challenges, opportunities, and implementation. *Proceedings of the IEEE* 102 (1): 56–103.

Felser, M. and Sauter, T. (2004). Standardization of industrial ethernet – the next battlefield? In: *Proc. IEEE Int. Workshop Factory Communication Systems (WFCS)*, September 2004, 413–420. https://doi.org/10.1109/WFCS.2004.1377762.

Gesbert, D., Shafi, M., Shiu, D.-s. et al. (2003). From theory to practice: an overview of MIMO space-time coded wireless systems. *IEEE Journal on Selected Areas in Communications* 21 (3): 281–302.

ODCA (2014). *Data Security Framework Rev 1.0.* Open Data Center Alliance (ODCA).

Osseiran, A., Boccardi, F., Braun, V., and Kusume, K. (2014). Scenarios for 5G mobile and wireless communications: the vision of the METIS project. *IEEE Communications Magazine* 52 (5): 26–35.

Prasad, R. (2013). Global ICT standardisation forum for India (GISFI) and 5G standardization. *Journal of ICT Standardization* 1 (2): 123–136.

Rusek, F., Persson, D., Lau, B.K. et al. (2013). Scaling up MIMO: opportunities and challenges with very large arrays. *IEEE Signal Processing Magazine* 30 (1): 40–60.

Saquib, N., Hossain, E., Le, L.B., and In Kim, D. (2012). Interference management in OFDMA femtocell networks: issues and approaches. *IEEE Wireless Communications* 19 (3): 86–95.

Zachary Zeltsan, 2003 Bell Laboratories, Lucent Technologies Rapporteur of Question 5 SG 17, ITU-T Recommendation X.805 Security Architecture for Systems Providing End-to-End Communications, Bell Laboratories https://www.coursehero.com/file/28313558/saag-3ppt/

Zhang, N., Cheng, N., Gamage, A.T. et al. (2015). Cloud assisted HetNets toward 5G wireless networks. *IEEE Communications Magazine* 53 (6): 59–65.

11

Issues and Challenges of 4G and 5G for PS

11.1   Introduction

Communication is one of the most integral components of human living. Not only is communication imperative in people's personal lives, but it has also been a critical component in the social and economic sector of every country and industry that operates globally. With the advent of technical means of connecting people around the world, there has been rapid growth in the means and methodologies of effective communication. One of the most popular is wireless connection of people in different locations. Technological advancement came along with a distinct improvement in communication devices and technologies. Compared to traditional methods of communication, people had to rely on wired devices such as telephones booths and landlines. Technical enhancements came along with wireless communication gadgets such as mobile phones, smartphones, tablets, smart television sets, and sensors, among many others. Such devices require unique technology that powers their communication capability. One of the most critical techniques was wireless broadband. Wireless broadband was a strategy created to create high-speed wireless connectivity among wireless devices at different locations for fast and reliable services and applications. Wireless broadband technologies were designed to enable public safety regarding interoperable communication, in addition to creating a valid and reliable wireless communication environment and platform.

Currently, people adopting different wireless broadband systems subscribe to specific service providers, who implement public safety policies for their users. Service providers have the sole responsibility of ensuring that their consumers have good and safe network connections, irrespective of their geographical position. Wireless broadband technologies have been advancing since their invention. For example, wireless connections have evolved from first generation (1G) to fifth generation (5G) approximately every decade. The latest developments regarding wireless connections are 4G, and for some countries, 5G networks. Despite their usefulness and economical means of connecting people around the world, they have been prone to various security breaches. Numerous public safety policies and organizations have been implemented to secure wireless connection, but there is still a vast gap to cover. This chapter takes a profound look at various issues regarding 4G and 5G network connections, their challenges, and various recommendations to mitigate the problems.

Public Safety Networks from LTE to 5G, First Edition. Abdulrahman Yarali.
© 2020 John Wiley & Sons Ltd. Published 2020 by John Wiley & Sons Ltd.

11.2 4G and 5G Wireless Connections

The 4G network connection has been one of the most popular in the last few years. It has been essential for many people regarding high-speed internet connections. On the other hand, 5G is still in its pilot phase of implementation in different parts of the world. Through various presentations and results from multiple experiments done by different stakeholders in the communication sector, it has been proven that 5G will be stronger and more efficient than 4G network connections. Before covering various public safety features in both network infrastructures, it is imperative to have an in-depth understanding of both 4G and 5G network connection technologies.

Various speculations began in the year 2008 as organizations had to develop multiple standards to oversee the development and implementation of the 4G network connections. 4G networks have multiple specifications that have made it accessible to many service providers and network users. First, the network infrastructure has the capability to adopt various network frequencies, where most of the frequencies are entirely dependent on the geographical location. For example, 4G network frequencies vary in different countries. For some countries, the frequencies vary in accordance with the town.

Secondly, 4G network connections have an extreme bandwidth that varies from 1 Mbps to 2 Gbps. According to the standards laid out by the International Telecommunication Union Radiocommunication Sector (ITU-R) organization, the network bandwidths are supposed to differ regarding facilities and equipment used in connections. For example, for heavy connections, the bandwidth should vary from 100 Mbps for high mobility connections (Thompson et al. 2014, p. 62–64.), which includes cars and trains, and 1 Gbps for low mobility connections, such as pedestrian and stationary devices. Such policies were implemented through the International Mobile Communication Advanced (IMT-Advanced) specifications. Other attributes of the 4G network include frequency band, which varies from 2 to 8 GHz. 4G networks are applicable and used in different network infrastructures such as wide area networks (WANs), local area networks (LANs), and metropolitan area networks (MANs), among many others. Moreover, the 4G networks have unified internet protocol (IP) addresses and seamless broadband utilization and integrations. Regarding service delivery, the 4G cellular network is supposed to provide an environment for dynamic data access, HD video streaming, and global roaming. Such services were rarely prevalent in previous mobile network technologies.

On the other hand, 5G is considered to be the future of internet connections. The internet has been a complicated technology that has been essential in making the world a smaller village. The high consumption demand has given rise to numerous research activities that have been instrumental in making network connections more efficient. Although it is still in development and in the pilot phase for some countries, the 5G network connection is set to succeed 4G networks. Just like 4G, 5G is also considered to be the fifth generation network connection. Regarding network connectivity, generation stands for the evolution of cellular networks since their invention and implementation. Fifth generation mobile connectivity is perceived to accommodate various characteristics. First, it is considered to have a bandwidth that is of higher speed than th 4G networks (Gopal and Kuppusamy 2015, p. 67–72). For example, it is supposed to accommodate speeds that are not less than 1 Gbps. Secondly, different from the frequency band evident in 4G connections, 5G is supposed to hold between 3 and 300 GHz.

Moreover, network infrastructure applicable to 5G cellular networks are almost similar to that of 4G connections, but with an addition of advanced technologies based on Orthogonal Frequency Digital Multiplexing (OFDM) modulation used in 5G. Furthermore, service delivery is also identical to that of 4G, only its connectivity will be based on a higher speed connection.

Based on the significant differences and character traits in both 4G and 5G cellular network connections, as stipulated above, it is evident that communication has been made more realistic, more comfortable, and more efficient. However, there have been continuous cases of security breaches during connections, especially for individuals connecting to the internet on a daily basis, and others using the internet as their primary mean of making a decent living. Some of these security breaches are conducted through hacking and cyberbullying, among others.

11.3 Public Safety for 5G and 4G Networks

Public safety can be defined as a network that is entirely dedicated to responding to various hazards that might be happening in society. The networking facility has been implemented since multiple governmental agencies responsible for public safety services for the community have had to communicate with one another during a hazardous event that requires attention from both government and the people in the community. Traditionally, public safety communications have concentrated on voice-based connections and narrowband communication infrastructures. With the advent of better technological means of increasing communication speeds, such users require better communication efficiency, which should be of better rate, to enable them to respond to various public safety reports that happen on a day-to-day basis. Although public safety organizations are still new to the new broadband technology, it is important for them to understand the implications of integrating the new technologies into their working environment, in particular having a better understanding on the differences between the latest two techniques of the 4G and 5G cellular network infrastructures.

Although safety does not only reflect on the government point of view regarding providing public safety services to its citizens, it is still a significant topic for both the private and governmental purposes. For that reason, there are some factors that have led to people and governmental agencies adopting the latest broadband cellular network facilities. First, broadband communications improve operational efficiency. Both private and governmental users have numerous operations going on at the same time. Some of such services include sending and receiving video transmissions, both low and high quality, database access, image transfers, and various office operations. Different from traditional network connections, broadband cellular networks have improved infrastructures that enable such services to be available for millions of people around the world with little or no latency.

The second reason for adopting broadband connections is the high demand for public safety. As mentioned earlier in this chapter, there have been numerous security breaches as far as the Internet of Things (IoT) is concerned, both at a personal level and at a governmental level. Considering some of the public safety concerns regarding network connectivity includes numerous terrorist communications and criminal activities online, which all fall under the government's agenda. With the advent of broadband communications, there are better standards that can control and inspect the flow of information from one point of communication to another. Finally, broadband connections are friendly economically. Compared with narrowband networks, which were more expensive regarding maintenance and technology, broadband connections have proved to be more pocket friendly. Broadband communications, through their high level of efficiency, enable consumers to come up with friendly consumption models, both for business purposes and also for private and governmental use. Despite their numerous advantages, as mentioned above, broadband communication infrastructures are prone to various security threats, which might breach public safety for its consumers.

11.4 Issues and Challenges Regarding 5G and 4G Cellular Connections

Threats that are related to 4G and 5G cellular networks can be put into various categories. Such categories cover both the internal and external component of any networking environment. Types of threat include attacks against privacy, attacks against integrity, attacks against authentication, and finally attacks against availability. The attacks can also be categorized regarding cellular network behavior that has either been reported by various consumers or through professional research activities.

11.5 Threats Against Privacy

Threats against privacy are widespread in society today. Such risks have been one of the primary reasons for various criminal activities that are prevalent in the community today. For example, it is common to get reports that are related to cyberstalking and bullying, where hackers gain access to private networks for their own gain. There are various types of attack that are associated with this category of threat. One of them includes eavesdropping attack, where an attacker intercepts a private network communication such as a phone call, video conferencing activity, fax communication, and or other types of connection that happen over a network. Another type of attack is known as a spoofing attack, where an attacker impersonates a legitimate network user to gain access to his or her private network for personal gain. Parallel session attack is yet another significant threat to privacy. In this type of risk, the attacker records a message from a previous IP communication and uses it in a current communication procedure for attack purposes. The final form of attack against privacy is known as replay attack. The attacker in this threat category uses valid information under transmission to maliciously repeat or slow down networking procedures for personal advantage or profit.

11.6 Threats Against Integrity

Threats against integrity have also been pervasive among various institutions. More specifically, such risks are common in business settings and various governmental and non-governmental institutions. They are threats that are focused on destroying the validity and integrity of a private or internal organizational information. For example, a bank cashier might change a single digit from a banking transaction, which in return jeopardizes the validity of the whole operation, which in return might give a false result at the end of a financial day. Similarly, threats against integrity can be put into different categories. One of them is a spam attack. Most internet users might receive unwanted advertisements either through emails or social media conversations (Ferrag et al. 2017, p. 7–14). Such frequent advertisements pose serious security concerns since they carry various threats such as trojan horses and other viruses that might affect an organization's or a private network. Another type of risk is known as a cloning attack. This type of attack is common among social media users. The attacker creates a fake account, with the name and details of their victim, and uses the fake account to gain more information from the victim's friends. The final mode of attack against integrity is message modification attack. The intruder, in this case, alters the packet header of a packet in transit on a network channel and directs it to a different destination. Alternatively, the intruder changes or modifies the message details from the recipient's side.

11.7   Threats Against Availability

From a telecommunications point of view, an attacks against availability have been common for various institutions. In particular, telecommunication companies and firms who are actively involved in the distribution of their products and services to their respective clients. The primary purpose of an attack against availability is to deny various individual accesses to various network services, for example, routing services. The threat is also known as a denial of service attack (DOS). Similar to other categories of attack, there are also different types of attacks under this current category. One of them includes a redirection attack. It is a type of attack that occurs when one is surfing online. The intruder redirects an online user to a different web page that is different from what they had requested. Redirection attacks are commonly associated with phishing attacks (Ferrag et al. 2017, p. 7–14). The second type of attack is known as a skimming attack. This type of attack is frequently associated with credit card services. For example, when a user is visiting a fuel station, the intruder might use a third party card reader that is different from the official fuel station card terminal. The intruder might use such a device to gain all information related to the cards, which includes the pin details. Such security threats are common in most business outlets.

11.8   Attacks Against Authentication

Attacks against authentication are an excellent example of an internal networking breach, which is also very common among users using the 4G services and previous forms of cellular networks. The primary objective of authentication attacks is to trigger server–client authentication and vice versa. There are various forms in which authentication attacks are practiced. One of them includes password reuse or password stealing attacks. This type of attack is practiced where an intruder gains access to a victim's password and uses it for personal gain. The second form of attack in this category includes the dictionary attack. This type of attack is evident where an intruder tries to use a different set of words to come up with an authentic password to gain access to a victim's network or personal computer.

11.9 Various Countermeasures to 4G and 5G Public Safety Threats

Although the threats discussed above have been of severe disadvantage to many people using broadband technologies, there are various methodologies that can be put in place to mitigate the issues. One of them includes cryptography methods. This is a methodology where the messages in a network are encrypted during transit and are only machine readable. The messages can only be decrypted if authorized personnel gain access to the message. The second countermeasure includes intrusion detection methods. Such a methodology is imperative for intruders that have entirely bypassed all the security measures and gain full identity on a network. Intrusion detection systems (IDS) are supposed to detect any unusual behavior once the intruder is in the network and alarm the necessary authorities. The third methodology involves the use of human factors. Human factors can be placed in three categories (Ferrag et al. 2017, p. 7–14). The first category involves what you know, i.e. strong passwords. The second human factor involves what you have, i.e. smart cards, pass cards, and tokens. The final human factor countermeasure involves who you are, which might include the use of fingerprints. Each of these factors has different drawbacks, which might include losing passwords and passcodes. However, it is the responsibility of a user to ensure that there is the maximum protection of such imperative human security accessories.

In conclusion, 4G and 5G networks prove to be of great importance as far as communication efficiency is concerned. However, most people and companies have reported losses of millions of dollars; others have claimed loss of essential security information that has downgraded their reputation and public image. Therefore, various service providers have the sole responsibility of ensuring that they have put up the necessary network infrastructure and policies that protect the public interest regarding public safety.

References

Ferrag, M., Maglaras, L., Argyrious, A. et al. (2018). Security for 4G and 5G cellular networks: a survey of existing authentication and privacy-preserving schemes. *Journal of Network and Computer Applications* 101: 55–82.

Gopal, B.G. and Kuppusamy, P.G. (2015). A comparative study on 4G and 5G technology for wireless applications. *Journal of Electronics and Communication Engineering (IOSR-JECE)* 10 (6): 67–72.

Thompson, J. et al. (2014). 5G wireless communication systems: prospects and challenges [Guest Editorial]. *IEEE Communications Magazine* 52 (2): 62–64.

12

Wireless Mesh Networking: A Key Solution for Rural and Public Safety Applications

12.1   Introduction

During emergency situations, reliable wireless mobile communications that enable real-time information sharing, constant availability, and interagency interoperability are imperative. One cannot depend on the existing infrastructure, but wireless mesh networking technology can be used effectively to provide broadband wireless communications for rescue operations. Given the shortcomings of current Public Safety and Disaster Recovery (PSDR), wireless mesh networks (WMNs) provide an interesting alternative technology. WMNs have been receiving a great deal of attention as a broadband access alternative for a wide range of markets, including those in the metro, emergency, public safety, carrier access, and residential sectors. This chapter provides the background and the challenges of wireless mesh technology. It addresses some of the technical influences of WMNs and, in particular, focuses on the opportunities that wireless mesh technologies provide for implementing an efficient network for emergency and public safety services.

Wireless technologies have played an important role in our modern world from simple voice and data communications for personal use to sophisticated communications for emergency and military organizations. In recent years, wireless broadband technology has rapidly become an established, global service required by a high percentage of the population. By the adaptation and integration of new technologies and applications, the wireless mobile industry is experiencing a revolution. Broadband wireless networks are turning into a strong foundation for digital information delivery that is changing the way we do business and lead our day to day lives.

There has been tremendous growth in both cellular as well as wireless local area networks (WLANs), which emphasizes that wireless communications without a doubt are a very desirable service. The current cellular networks and even the third generation (3G) of cellular networks such as Enhanced Data for a Global system for mobile communications (GSM) Evolution (EDGE) and universal mobile telecommunications systems (UMTSs) offer wide area coverage, but the service is relatively expensive and offers low data rates when compared to WLAN 802.11 (Yarali, 2008). However, WLANs have limited coverage and mobility, and a wired backbone connecting the multiple access points is required to increase the coverage areas of WLANs.

Wireless metropolitan area networks (WMANs) provides higher data rates and also a very good quality of service to a relatively large customer base (Yarali, 2008). WMANs

Public Safety Networks from LTE to 5G, First Edition. Abdulrahman Yarali.
© 2020 John Wiley & Sons Ltd. Published 2020 by John Wiley & Sons Ltd.

require line of sight and they also do not support mobility. The wireless mesh networks almost eliminate these disadvantages.

WMNs offer low cost and high-speed broadband internet access for both fixed and mobile users. Some of the advantages of WMNs over WLANs are: they require few access points, no need of line of sight, have weatherproofing, fast deployment, high outdoor scalability, less complex, low backhaul cost, and low installation costs. Wireless mesh networks are next generation wireless networking technology that has evolved recently. In wireless mesh networks, the nodes include mesh routers and mesh clients. Each node acts like a host as well as router (Akyildiz et al., 2004). Every node forwards packets of information on behalf of other nodes that may not be within direct transmission range of their destinations. Self-configuring and self-organizing are the two main features of wireless mesh networks. With these two features, nodes can automatically maintain mesh connectivity among themselves. Low upfront cost, easy network maintenance, reliable service coverage, and robustness are some of the benefits drawn from the self-configuring and self-organizing feature of wireless mesh networks.

In a crisis situation, providing emergency services with accurate and up to date information is very important. One cannot depend on the existing network infrastructure to establish communication between rescue workers during emergency situations. During disasters and emergency situations, there is a need for a low cost, fast deployable, secure, redundant broadband communication network. Wireless mesh networks have all the features required to set up a broadband communication network during such scenarios.

During disasters and emergency situations, wireless mesh networks can provide connectivity to emergency vehicles anywhere in the city/town, which helps police and fire departments to serve their community better. Wireless mesh networks improve communications and provide instant access to local and national database systems at a low cost (White paper, n.d.). Video surveillance, mobile, and temporary networks for emergency situations, personnel, and vehicle tracking are some other public safety applications that benefit from wireless mesh networks (White paper, n.d.).

12.2 Wireless Mesh Networks

In a traditional wireless network where a mobile device such as a laptop connects to a single access point, each laptop has to share a fixed pool of bandwidth. With mesh technology and adaptive radio, devices in a mesh network will only connect with other devices that are in the set range. The more devices, the more bandwidth becomes available, provided that the number of hops in the average communications path is relatively low (Yarali, 2008). Reliability is enhanced by defining multiple paths through the mesh. Capacity can be enhanced if these paths are implemented with multiple radios. Figure 12.1 is an example of mesh network configuration.

WMNs have several favorable characteristics, such as dynamic self-organization, self-configuration and self-healing, easy maintenance, high scalability, and reliable services. All these favorable characteristics make WMNs a cost-effective way to enable police and fire departments and other first responders to handle disasters and emergency situations by providing connectivity to emergency vehicles anywhere in the city or town. Video surveillance, mobile and temporary networks for emergencies, and

Figure 12.1 Mesh network topology.
Source: Farpoint Group.

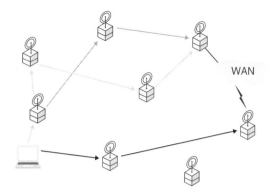

WAN

personal and vehicle tracking are some other public safety applications that benefit from WMNs.

To improve network performance, mesh networks can enable users to share bandwidth and also balance the load on the network. They are a much cheaper, easily deployable, less expensive alternative to wired access points. Adding access points does not incur many costs of adding more wire. The range and coverage of the network can be improved by adding wireless access points. Wireless nodes should be self-configuring, self-healing, and self-tuning in order to ensure a scalable wireless mesh network. This makes it possible to deploy extremely large scale, intelligent, high-speed networks with predictable performance and resiliency. The mesh eliminates the bottlenecks that might occur in single-hop networks by balancing the load on the network.

There are several factors that have to be considered for the successful deployment of WMNs. Optimal performance can be achieved only if the WMNs are carefully architected. The first key for the successful deployment of WMNs is having the appropriate MM router density. The node density can neither be chosen too low nor can it be chosen too high. The second key to successful deployment is sufficient and appropriately placed mounting assets. The third key is to distribute and strategically place backhaul in order to meet end-user needs. The fourth key is to plan back-end integration in order to eliminate any operational difficulties that might occur if back-end integration with the wired network is neither properly considered nor properly implemented. The fifth factor to be considered is network-wide management by using sophisticated network management tools (Yarali, 2008). WMNs deployment can be made incredibly fast, low cost, and simple by proper planning and consideration.

12.3 WMN Challenges

Some major issues that pertain to WMNs are capacity and interference. In mesh architectures, capacity and interference are closely linked. Issues that affect efficiency can be combated with the use of antennas.

Non-directional antennas have been deployed in mesh systems because in general they are small, unobtrusive, can link more than one other node and require no alignment. Reducing the transmitting power may lead to reductions in power consumption that could potentially benefit portable battery-driven devices. In fixed link services, this is not significant because in modern broadband wireless systems the power consumption is often dominated by signal processing rather than RF circuits (Yarali, 2008). Also, the transmitter normally operates when there is traffic to pass and may only represent a small portion of the power budget. Therefore, the main advantage of using a directional antenna at the transmitter is the reduction of interference caused to victim receivers outside the main beam. Inside the main beam, interference is no worse than for Omni systems.

Besides using antennas to improve the efficiency of the wireless mesh, we must also consider the channels through which wireless networks operate. Despite the availability of multiple non-overlapping channels in the 2.4 GHz and the 5 GHz spectrum, most IEEE 802.11 based multi-hop and ad hoc networks operate in only a single channel. This reduces the available bandwidth to each node due to the interference between successive hops on the same path as well as between neighboring paths. As a result, single channel ad hoc network architecture cannot be applied to form effective WMNs that can adequately support the bandwidth requirements of the last mile wireless broadband access network, or campus scale wireless backbone. However, a novel multi-channel WMN architecture can effectively address this bandwidth problem by fully exploiting the non-overlapped radio channels made available by the IEEE 802.11 standards (Yarali, 2008).

12.4 WMNs for Disaster Recovery and Emergency Services

There are two worst-case scenarios for emergency response personnel where they must bring their own, complete communication capability with them. The first is the natural or manmade disaster scenario where all land-based, cellular, Wi-Fi, and WiMax infrastructure is destroyed or disabled. The second is the remote disaster site that has no terrestrial communications infrastructure. The second case is a much more common situation than many people would expect as there are still vast stretches of rural and remote areas in the United States that are not supported by terrestrial commercial communications.

The conventional solution for communication in disaster relief operations is largely based on Terrestrial Trunk Radio (TETRA), designed for speech and status messaging, with data rates between 2.4 and 7.2 Kbps (De Graaf et al., 2007). These data rates are insufficient for public safety officials or the fire services to access details of the disaster site or to transmit or receive video images of the disaster site. Events such as 9/11 or Hurricane Katrina have demonstrated that there exist significant inadequacies in first responder communications. One of the main problems the rescue teams and emergency services faced during these disasters was the lack of interoperability between communications equipment used by different public safety agencies and jurisdictions (Portmann & Amir Pirzada, n.d.). Another problem with PSDR communications was the strong reliance on terrestrial communications infrastructures such as traditional landline and cellular telephony as well as infrastructure based land mobile radio.

To overcome such problems, deployment of high bandwidth, robust, self-organizing, self-healing and self-configuring, low cost, and reliable wireless mesh networks are preferred over the more vulnerable low bandwidth networks. WMNs do not depend on a wired backbone infrastructure. WMNs are comprised of wireless mesh routers that provide network access to wireless clients. Typically, communications between these mesh routers take place via the wireless network involving multiple wireless hops. One or multiple mesh routers that are connected to the internet can then serve as gateways for all other nodes and provide internet connectivity to the entire mesh network (Portmann & Amir Pirzada, n.d.). WMNs allow relief workers to communicate with each other at the disaster site more quickly and also in a cost-effective way.

12.5 Reliability of Wireless Mesh Networks

Reliability is the fundamental thing to be considered for the successful deployment of networks during emergency and disaster situations. First of all, the system architecture should be such that the failure of the network elements within the network should not have an impact on large parts of the network. Secondly, if there are any disruptions in the communications path, they have to be detected automatically. The communication topology should have an alternative communications path and reconfiguration of the paths should take place. Thirdly, the communications equipment must be resistant to extreme temperatures, shock, and vibration, salt and dust, radiation, rain snow, etc. (De Graaf et al., 2007).

The WMN has the ability to reorganize itself and keep functioning even if one or more nodes are moved from one location to another, or simply removed from the network. This function is made possible because of the redundancy existing in the mesh topology. If a node in a mesh network fails, messages are sent around it via other nodes and the operations of the mesh networks are not affected by the loss of one or more nodes. The WMNs support multi-hop routing in which there is more than one path from an end node to another. Thus the degree of route redundancy enhances the mesh network reliability (Wireless Mesh Networks Reliability and Flexibility, n.d.). WMNs are able to provide service even during weather conditions like extreme temperatures, shocks, and vibrations, salt and dust, rain and snow.

12.5.1 Self-configuration of Wireless Mesh Networks

During disasters and emergency situations, the networks should be deployed easily and quickly and with little human maintenance. Devices that have the capability to configure themselves are preferred during such situations. The WMN has the ability to build and configure itself. Whenever each end node is powered on, it searches for the neighboring nodes and if it finds one or more, it issues a request to join the network. It is admitted to the network only if it meets the admission criteria set at the security settings. Once the end node joins the network, it does not need human intervention to send a message to the destination. Paths will be automatically formed by the end node as the information it transmits gets relayed by neighboring nodes until it reaches the central node (Wireless Mesh Networks Reliability and Flexibility, n.d.). WMNs are also self-reconfiguring networks. Intelligent self-reconfiguration also makes the mesh self-tuning end to end. This feature allows traffic to move dynamically along the optimal paths. Traffic is sent based on the shortest or fastest or least congested or other characteristics of each path (White paper, n.d.).

12.5.2 Fast Deployment and Low Installation Costs of Wireless Mesh Networks

As discussed earlier, WMNs have the ability to build and configure themselves. The WMNs are also capable of functioning even if one or more nodes are moved from one location to another or simply removed from the network. These self-configuring and self-healing capabilities of WMNs make them a fast deployable network that is necessary for emergency and public safety operations. Several companies have already realized the potential of this technology and have begun to provide fast deployable WMN equipment. New Energy Technologies, a Chicago based company is offering a wireless mesh "smarter kit" which can deliver one square mile of wireless communications. The "smarter kits" can be deployed from scratch in less than an hour. The "smarter kits" are fast deployable mesh kits that support internet, voice over internet protocol (VoIP), video, and data communications using carrier-grade wireless broadband technology supplied by Ontario based BelAir Networks (White Paper, n.d.).

WMNs eliminate the need for expensive cabling thereby reducing implementation costs. The self-configuration feature of WMNs allows installation to occur in hours instead of days or weeks thereby reducing labor-intensive deployment costs.

12.5.3   Voice Support of Wireless Mesh Networks

WMNs have the ability to support voice transmissions, which helps public safety and emergency response officials to communicate with each other and also with the central office effectively. The public safety and emergency response communications can be improved with the use of Wi-Fi-based voice devices such as dual-mode cellular/Wi-Fi and pure Wi-Fi based handsets, laptops, and smartphone voice clients (White Paper, n.d.). Some of the major benefits include low voice degradation, high clarity in voice, and reduced voice costs.

12.6   Video/Image Support of Wireless Mesh Networks for Emergency Situations and Public Safety

During disasters, real-time video communications will significantly aid the ability to describe the disaster scene for first responders and public safety officials. Deploying a traditional wire-based system for video applications during such a scenario is expensive and also time-consuming. Instead, deploying WMNs is a cost-effective and quick alternative. Other benefits of deploying the WMNs include increased safety and more efficient traffic flow.

12.6.1   Video/Image Support of WMNs for Large Disasters

During a large-scale disaster, it is reasonable to create a MANET for a team of rescue workers working at the particular disaster site. Typically, MANETs are based on 802.11 b and/or g Wi-Fi and may span to few kilometers with multi-hop mesh routing (Kanchanaburi et al., n.d.). Figure 12.2 illustrates the WMNs for multimedia disaster emergency response communications.

As shown in the figure above, a local MANET that runs on Wi-Fi is formed at each disaster site using laptops and PDAs. A special node called a WMR that has both Wi-Fi and satellite interfaces is installed at each site. The WMRs allow the traffic to pass among the disaster sites and the command headquarter. To hide the network heterogeneity a virtual private network can be created among the WMRs and the mesh router at the command

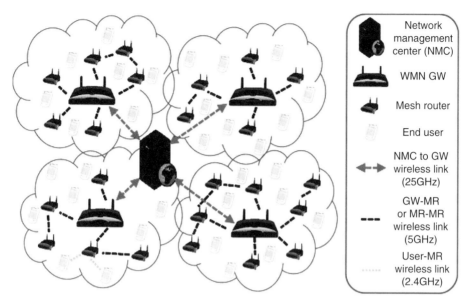

Figure 12.2 WMNs for multimedia emergency response communications during large disasters [https://link.springer.com/article/10.1007/s11277-013-1579-9].

headquarters. An Optimized Link State Routing protocol (OLSR) can be used as a WMN routing protocol to allow IP based multimedia communications between the command headquarters and the disaster site (Kanchanaburi et al., n.d.). Three application software components can be extremely helpful during emergency situations. The first one is multimedia communications software through which every rescuer can make voice, video calls, and also instant messaging to any other rescuer and also with the headquarters. The multimedia communications software runs on laptops with a built-in web camera, microphone, speaker, and Wi-Fi. The rescuers can easily carry such laptops to the disaster site and facilitate communications. The second application component are sensors that can be integrated into WMNs to measure temperature, humidity, wind speed, wind direction, and rainfall. This can be extremely helpful for the planning of rescue operations. The third application component is face recognition software that can be used to compare faces captured at the disaster site by the rescue workers with a collection of known faces. Rescuers can send live images from the disaster site through the WMNs to the command headquarters where the matching of images takes place (Kanchanaburi et al., n.d.).

12.6.2 WMNs Supporting Video Monitoring for Public Safety

As shown in Figure 12.3, WMN deployment can be used by public safety or other security agencies for video surveillance of areas where the crime rate is higher in highway corridors or downtown. Security monitoring is very helpful for public safety. WMN deployment is one of the cost-effective ways to achieve the video monitoring capability of the network. In this deployment, the MPEG4 video stream is transmitted through the access points across the backhaul connection to the core network and to the control

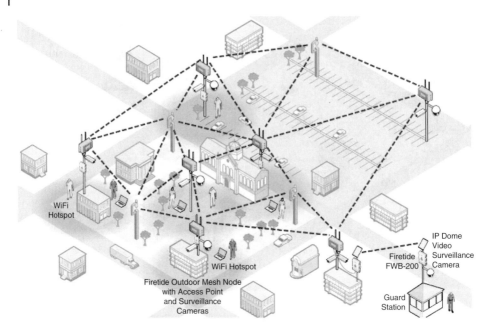

Figure 12.3 WMNs supporting video monitoring of urban centers, traffic corridors, and highways [http://europeindustrynews.com/23126/enterprise-wlan-market-to-experience-increase-in-growth-by-2016-2024].

center (White paper, n.d.). The suspicious behavior and unsafe conditions are monitored by the operators at the control center.

12.6.3 WMNs for Mobile Video Applications of Public Safety and Law Enforcement

WMNs can also support mission-critical mobile video applications for public safety and law enforcement. Mobile video can be used by police departments and fire departments. As shown in Figure 12.3, wireless cameras can be installed on the public safety vehicles and ambulances and can be used to obtain real-time video coverage of events such as high-speed pursuits, wildfires, or triage examinations of victims as they are being rushed to the medical facility (White paper, n.d.). In addition to this, in order to receive video from other stationary or mobile sources, emergency vehicles can be equipped with mobile gateways. The mobile video support can better prepare the public safety officials to face the situation they might encounter at the scene of the incident.

12.7 Interoperability of WMNs for Emergency Response and Public Safety Applications

According to SAFECOM, the department of the homeland security communications program, interoperability refers to the ability of emergency responders to work seamlessly with other systems or products without any special effort. Wireless communications interoperability refers to the ability of emergency response officials to share

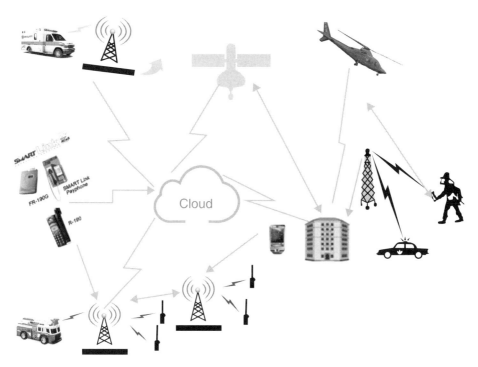

Figure 12.4 WMNs supporting mission-critical mobile video applications for public safety and law enforcement [https://hal.inria.fr/hal-01400746/file/sdn-public-safety%286%29.pdf].

information via voice and data signals on demand, in real time, when needed, and as authorized (White paper, n.d.).

The need for accessing and sharing data and images dictates the need for interoperable broadband wireless mesh networks that can connect highly mobile workers, deliver high bandwidth of data, are highly reliable, and support real-time voice, video and data, as depicted in Figure 12.4. As shown in the figure, WMNs can support GIS mapping file (huge files) data transfer, IP cameras for real-time video surveillance of the disaster site, and monitoring capabilities at the command centers and at the emergency operations centers.

In addition to this, WMNs also provide reliable communications, high security, quality of service (QoS) and performance that are highly essential during emergency situations.

Another key element of the mission-critical interoperable wireless network is the ability to support instant expansion and quickly scale to support many new users. The WMNs can be expanded easily and incrementally as needed. For example, in the case of the Minneapolis bridge disaster, the WMNs were fully expanded within hours of a disaster to provide full coverage of the disaster site, supported the addition of real-time video surveillance, and were scaled to address 6000 concurrent users (White paper, n.d.).

12.8 Security in Wireless Mesh Networks

Security is one of the major concerns, especially for a public safety and emergency services wireless network. Data transmitted on the network should be protected from external intrusions and unauthorized access to the WMNs. Digital signatures are one of the best features that can be used to ensure that only authorized systems are allowed to access the mesh network. Wireless intrusion detection sensors can be used to detect common 802.11 attacks, which include MAC management attacks as well as rogue access points attacks (Gerkis & Purcel, 2006). User traffic between the nodes can be encrypted to prevent wireless eavesdropping. Industry standards such as 128 bit and 256 bit AES encryption can provide very effective security comparable to the level of security provided by wired networks (White paper, n.d.). WMNs can also be secured with the combination of WLAN, virtual private networks, firewall, intrusion detection controls, and application controls such as HTTPs (Gerkis & Purcel, 2006).

12.9 Conclusion

During disasters, one cannot depend on the existing infrastructure for establishing communication between the rescue workers. There is a need for a low cost, quickly deployable, secure, redundant broadband communication network to handle such situations.

WMNs have the ability to reorganize themselves and keep functioning even if one or more nodes are moved from one location to another, or simply removed from the network. This makes the WMNs more reliable. WMNs have the ability to build and configure themselves. These self-configuring and self-healing capabilities of WMNs make them a quickly deployable network. WMNs eliminate the need for expensive cabling thereby reducing the implementation costs. The self-configuration feature of WMNs allows installation to occur in hours instead of days or weeks thereby reducing labor-intensive deployment costs. WMNs can support both stationary as well as mobile video applications of public safety and emergency responders. WMNs can also support voice transmission on the network, which helps public safety officials and also first responders to communicate with each other effectively. WMNs can connect highly mobile workers, deliver high bandwidth of data, are highly reliable, and support real-time voice, video, and data. WMNs support interoperability, which is highly essential during emergency situations. By using digital signatures and wireless intrusion detection sensors WMNs can be made more secure.

Reliability, self-organizing, self-healing and self-configuring, fast deployment, support for stationary and mobile video and voice applications, and the security of WMNs makes them strongly recommended for deployment during emergency and disaster situations.

References

Akyildiz, I.F., Wang, X., and Wang, W. (2004). *Wireless Mesh Networks: A Survey*. Elsevier.

De Graaf, M., Van Den Berg, H., Boucherie, R.J. et al. (2007). *Easy Wireless: Broadband ad-hoc Networking for Emergency Services*", Med Hoc Net, 32–39. Corfu Greece: Ionian Academy.

Gerkis, A. and Purcel, J. (2006). *A Survey of Wireless Mesh Networking Technologies and Threats*. GIAC Gold Paper.

Kanchanaburi, K., Tunpan, A., Awal, M.A. et al. (n.d.). *Building a Long Distance Multimedia Wireless Mesh Network for collaborative Disaster Emergency Responses*. Thailand: Internet Education and Research Laboratory, Asian Institute of Technology.

Portmann, M. and Amir Pirzada, A. (2006). Wireless mesh networks for public safety and disaster recovery communications. In: *Chapter in Wireless Mesh Networking: Architecture, Protocols, and Standards Auerbach Publications*, 546–573. CRC Press.

White paper (2004). *An Introduction to Wireless Networking*. Los Gatos, CA: Firetide.

White Paper (2008). *Mesh Kits for emergency WLAN and 802.11n for Interop show*. Edinburgh: Compute Scotland.

White Paper (2006). *Solution Brief: Wireless Mesh Network*. Nortel.

White paper (2006). *Delivering Video Over Mesh Networks*. Belair Networks.

White paper (n.d.). *Public safety Interoperability: Wireless Mesh Delivers*. Belair Networks.

Krishnaiah, R.K., Ramesh, B.B., Kumar, P.T. et al. (2009). Wireless Mesh Networks Reliability and Flexibility. *Oriental Journal of Computer Science and Technology*: 23–29.

Yarali, A. (2008). "Wireless Mesh Networking Technology for Commercial and Industrial Customers". IEEE 21st CCECE08, Niagara Falls, Canada, 47–62.

13

Satellite for Public Safety and Emergency Communications

13.1 Introduction

Satellite communication in the twenty-first century has gained a place in the spectrum of communication approaches, serving multiple purposes, even in the presence of advances such as fiber optics. Globally, investments of over $100 billion characterize these systems offering a supportive infrastructure for communications in multiple governments and businesses (Elbert 2008). The systems conventionally apply a satellite in a geostationary earth orbit geosynchronous satellite system (GEO) rotating around the equator, or any other fixed point, once in 24 hours and maintaining synchronicity with the movement of the Earth (Elbert 2008). The GEO satellites are located approximately 36 000 km from the surface of the earth, offering global coverage at once despite the two-second possible delays in communication emanating from the distance (Evans et al. 2005). There are non-GEO satellites located less than 36 000 km from the surface of the earth, which offer fewer delays during the propagation of communication (Elbert 2008). Nevertheless, the latter forms have limited coverage for communication.

Satellite communication has found a popular market in the area of public safety communication. The application of these systems proved beneficial in the earthquake disaster of 1985 in Mexico City when satellite communication persisted despite the failure of all other terrestrial communication forms. (Elbert 2008) In the twenty-first century, the restoration of communication after the Sumatra earthquake depended on these communications (Elbert 2008). Satellite communication exhibits better performance than cellular communication because satellite remains unaffected by cellular dead zones or the occurrence of a temporary environmental condition like floods or earthquakes owing to the location of the satellite in space (Ground Control 2019). It is also not subject to failure due to weather conditions or human interference, as is the case with terrestrial systems of communication.

As such, the viability of satellite communication as a framework for public safety communication is a significant area of interest. It presents the possibility of more stability in the maintenance of communication and potentially more effectiveness in coordinating emergencies. The understanding of its capacities, its popularity both nationally and globally, as well as the potential forms of communication support it presents, should be instrumental to efficiency in the future management of public safety.

13.2 Contextualizing Public Safety

There are multiple entities involved in operations deemed as part of the pursuit of public safety. While the range of actions is vast and tends to vary based on specific occurrences, it is possible to classify and delimit the nature of public safety operations and organizations. One of the main areas of public safety is law enforcement. This refers to the prevention, investigation, apprehension, and detention of individuals suspected or convicted of offenses that are against the law (Baldini et al. 2014). Public safety also features emergency medical and health services (EMHS), which is the provision of critical care for the sick and injured as well as the transfer of individuals to safe and controlled environments. The conventional providers of these services include doctors, nurses, paramedics, and trained volunteers (Kumbhar et al. 2016).

Additional functions in the area of public safety cover border security, including activities executed by border security, the navy, or any specialized police forces. Firefighting and environmental protection activities also feature within the scope of public safety operations (Baldini et al. 2014). Firefighting may take in the context of buildings or the urban environment, as well as situations like wildfires in forests or grasslands. The range of environmental protection actions is vast, but the most conventional include the protection of the ecosystem in either specific regions or nationally (Baldini et al. 2014). The activities may feature monitoring of water sources and catchment areas, air pollution, and land monitoring.

Public safety also involves search and rescue. These are operations involving the location and transportation of lost or missing people from their danger to a place of safety. Various safety organizations are also involved in processes of crisis management (Kumbhar et al. 2016). Emergency crises involve the creation of multiple supply chains featuring civil engineering as well as various bodies that handle security, health, and search and rescue. It is notable that while some of the conventional functions in public safety may be limited to the regional level, most of them pursue operations at the national level (Baldini et al. 2014). This has results such as otherwise independent organizations often requiring collaborations at the national level, such as local search and rescue operations being facilitated by the police or several firefighting organizations (Baldini et al. 2014). The perspective contextualizes the need for efficient public safety communications and the exploration of satellite communication as an option toward effectiveness.

13.3 Public Safety Communications Today

The design of public safety communications is expected to exhibit levels of efficiency that ensure the safety of both the responders and the civilians experiencing emergencies. Indications are that the public safety community currently is dependent on land mobile radio (LMR) systems for the provision of support and mission-critical voice communications (Department of Homeland Security 2014). The use of radio systems has evolved from the traditional application of private systems to modern antiquated systems with or without vendor support (Desourdis et al. 2001). The systems have been relatively reliable for the provision of means for personnel in the field to communicate with each other, making extensive use of public safety answering points (PSAPs) as well as the supporting communications centers.

Nevertheless, it is notable that with the evolution of LMR systems, there have been several different systems that have emerged on a national basis. The core problem with these systems is the potential incompatibility of the systems in use nationwide. The outcome is a constant struggle with interoperability, which often implies the diminished capacity to facilitate communications across jurisdictional and agency lines (Carl et al. 2016). The situation exacerbates as some public safety agencies continue the application of combined low bandwidth LMR and high-speed broadband data services (Ferrus et al. 2015). The latter derives from commercial networks and is often applied in the support of response efforts as well as the performance of wireless data access functions.

The functions include the execution of license plate searches, messaging, sharing low-resolution messages, and dispatch (Desourdis et al. 2001). These functions are all critical to communication efforts, but the majority of broadband data solutions exhibit the limitations concerning the capacity for interoperable support of emergency responses and operations. Other conventional problems relating to the wireless broadband solutions include the failure to apply public safety standards and practices to enhance resiliency and robustness (Department of Homeland Security 2014). They may also present accessibility challenges due to the absence of the availability of network components. Therefore, the use of LMR in public safety communication presents challenges in increased costs for acquisition, and the differences in sourcing and application also poses issues in interoperability. These aspects create the need for equally beneficial but similar systems that are applicable in the context of public communications for emergencies and critical contexts.

13.4 Satellite Communications in Public Safety

The efforts in addressing public safety communications span several alternatives, one of them being the application of satellite communications. This is the core focus of this report, seeking to contextualize the nature of satellite communication, its merits, and the potential shortcomings it could present within the public safety framework.

Communication satellites refer to microwave repeater stations that facilitate communication through the reception and changes to transmitted signals from one point of the Earth through space to another point. Satellite communication is among the earliest advances in wireless technology, presenting advantages including broadcasting possibilities as well as coverage of extensive spaces and distances (Elbert 2008). The principles along which the functionality of the satellite is achieved are expressed in Kepler's Laws. The first law states that every planet revolves around the sun in an elliptical orbit and the sun is one of its points of focus (Maral and Bousquet 2011). The law implies that the earth is one of the foci for the satellite as it continues on its orbit in space, forming the basis for the entire concept of revolving satellites.

The second law states that within similar amounts of time, the distance covered by the satellite is equal relative to the center of the Earth (Hussaini 2017). This law ensures that transmission remains constant, and handover among satellites is unaffected despite the potential differences in the point of the Earth over which the movement occurs. The third principle attests to the square of the periodic orbit time being proportional to the mean distance of the two bodies to the power of three (Maral and Bousquet 2011). The positioning of different satellites in space, therefore, can be determined using this law.

The orbital period of a satellite is only dependent on its distance from the Earth and no other factors.

13.4.1 Topology and Frequency Allocation

Satellite networks have multiple characteristics, one of the most essential being their network topology. One of the topologies, the meshed network topology, involves the presence of nodes where every single node can communicate with every other node (Maral and Bousquet 2011). The meshed satellite network consists of multiple earth stations that communicate with each other through satellite links with radio frequency carriers. Conventionally, this network applies either the transparent or regenerative satellite (Elbert 2008). However, the former requires that the quality of the radio frequency link between two earth stations be sufficiently high for efficient service (Maral and Bousquet 2011). Figure 13.1 is an example of the star topology of satellite connections.

The star network topology features each node communicating using one central node or the hub. It is possible to have multi-star topologies, featuring several hubs that allow the satellite communications. In the context of communication, the hub is a large earth station whose antenna size may be as large as 10 m (Maral and Bousquet 2011). The earth station has a higher equivalent isotropically radiated power (EIRP) than the other earth stations within the network, giving it higher transmission integrity than the other nodes (Basari et al. 2010). This topology benefits from the communication with the hub due to the minimal challenges related to the EIRP. In networks applying the very small aperture terminals (VSATs), the architecture is immensely popular due to the benefits of the hub (Maral and Bousquet 2011). The small terminals connect to the hub using the inbound link while the hub links to other stations using the outbound link.

13.4.2 Satellite Communications

Undoubtedly, satellite communication for public safety has been consistent in areas such as the military and emergency care provision. While the high cost of the dedicated satellite has proven prohibitive toward application in some safety contexts, there are specific communication forms that are consistently beneficial toward the transmission of urgent information (Maral and Bousquet 2011). In telemedicine, for instance, the use of

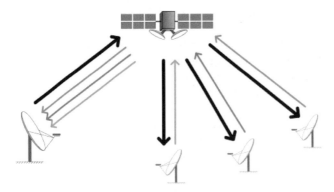

Figure 13.1 The star network topology. Source: http://www.servicesat.net/private_networks.php.

satellite technology has proven effective for emergency responses especially in isolated areas with limited terrestrial coverage (Latifi 2011). Military operations have also long made use of GPS in field operations (Baldini et al. 2014). Portable satellite technologies have become available. Some of the communication resources include the broadband global area network (BGAN), which provided the first mobile voice and data communications through the application of portable terminals (Richharia 2017). The technology has capacities for download speeds reaching 492 kbps, which beats conventional satellite communication (Latifi 2011).

The VSAT is another form of satellite connectivity that facilitates the transmission of broadband and narrowband data (Urgent Communications 2010). VSAT is highly portable and their design typically suits real-time data transmission. While they are reliable, their applicability to the transmission of bandwidths as with other satellite options remains limited. However, they prove valuable in remote areas where conventional transmission or communication may have undergone damage, such as in the course of rescue operations during natural disasters. VSAT is typically more affordable than the BGAN, but it is difficult to conduct transmissions such as video content on the network due to the limitations of bandwidth (Latifi 2011).

Therefore, satellite communication already proves viable for the conduct of communication, especially in areas unreachable by typical forms of technologies, or where these technologies have been disrupted. Nevertheless, harnessing this communication technology toward fully benefiting the public safety goals requires an in-depth exploration of its potential for application and areas of failure.

13.4.3   Applications of LEO and GEO Satellites in Public Safety Communication

While satellite communication for public safety was once considered a high expense venture, progress in the area has facilitated the accomplishment of this form of communication as a viable alternative for multiple agencies (Urgent Communications 2010). This viability stems from the reduction in size and weight toward the accomplishment of affordability due to low upfront cost. Similarly, there has been the growth of possible satellite devices including the emergence of communication devices such as fax, dispatch radios, and data tracking capacities (Desourdis et al. 2015). Therefore, there are multiple possible applications of satellite technologies in the context of public safety communication, featuring satellites at different locations from the surface of the Earth.

The difference in the location in terms of the distance from the surface of the Earth often, but not always, coincides with the size of the satellite. They include low Earth, medium, and geosynchronous orbits (Figure 13.2). Low Earth orbits satellite systems (LEOs), for instance, are often small and exhibit low power requirements (Urgent Communications 2010). The satellites are located close to the surface of the Earth, in particular due to concerns regarding collisions with debris in outer space (about 1000 mile from the Earth's surface). Their size tends to limit the capacity to track them, which escalates the possibility of unprecedented collisions in space (Desourdis et al. 2001). These satellites, therefore, owing to the low course over the Earth, complete their orbit rapidly. Their proximity to the earth diminishes their power needs for maintenance and the mobile transceiver requires only a small omnidirectional antenna (Urgent Communications 2010).

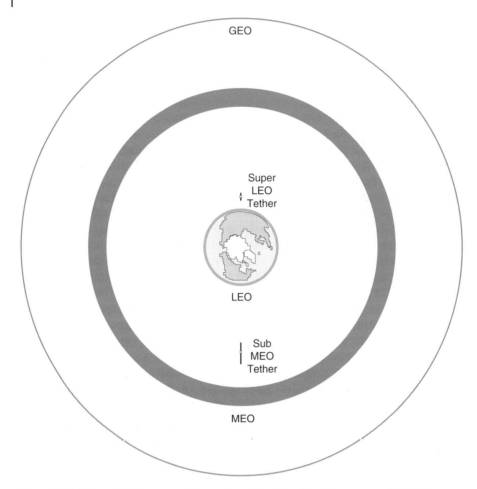

Figure 13.2 LEO and GEO in context. Source: http://hopsblog-hop.blogspot.com/2016/08/tran-cislunar-railroad.html.

Notably, however, there are big and mega-LEOs applied for satellite communication services. These LEOs have higher bandwidths, with the big LEOs operating above 2 GHz and the mega-LEOs functioning in the range between 20 and 30 GHz. Therefore, the mega-LEOs will often have the capacity to carry higher information loads, which makes them suitable for the transmission of video content with minimal delay.

Due to these features, the small satellites applied for safety communications exhibit lower cost of construction and maintenance. However, they also produce difficulties in sustaining communication due to the need for more satellites in the case of the intention to provide global coverage (Holmstrom et al. 2011). At the same time, they exhibit more complexity due to the need for constant handoff of calls among the satellites that are always in rapid movement. Similarly, the intention to use this form of satellites demands the setting up of multiple infrastructures on the ground for effectiveness. Systems such as Iridium and Globalstar have characterized these low orbit small satellites in application within public safety scenarios (Desourdis et al. 2001). Iridium operates with 66 LEO satellites, allowing the provision of global coverage through access to the satellites especially in the absence of cellular coverage.

GEOs are much larger and usually require more advanced equipment for their launch into space. Located more than 36 000 km from the earth, GEO systems work on the premise that the placement of a satellite above the Equator with considerations for altitude will experience orbiting speeds that match those of the Earth's rotation (Urgent Communications 2010). The implication of this assumption is that the position of the satellite in the sky remains the same as a given fixed point on the Earth's surface, allowing continued transmission (Holmstrom et al. 2011). For systems such as fixed dispatch locations in the public safety scenario, the application has an obvious advantage. Only a few satellites are necessary for the provision of global coverage and service. Nevertheless, the distance from the surface of the Earth and the large size of these satellites implies the higher application of power as well as escalation of the upfront and maintenance costs. The conventional application of these systems is in areas such as weather forecasting and satellite radio communications.

The nature of satellite communication systems in public safety features the integrated availability of fax, packet-switched data, and push-to-talk (PTT) two-way dispatch radio, as well as telephony and circuit-switched data (Urgent Communications 2010). Application of both GEO and LEO communications in public safety, especially in the rural areas, specifically benefit from the satellite capacities of the dispatch radio. Switching to a satellite system has already proven beneficial in multiple rural contexts where cellular communication proves inherently impossible. This is exemplified by the case of Emergystat and its deployment of the NetRadio system in Vernon (Urgent Communications 2010). The effort featured the installation of satellite terminals on 15 emergency response vehicles as well as another fixed one at the dispatch center. The analysis of this system indicated that the units connected in a single dedicated satellite connection, with the capacity to save about $10 000 annually, from the use of the satellite dispatch radio system (Urgent Communications 2010).

Indications are that while the benefits of satellite systems in rural areas are evident, they are also applicable to the less obvious urban scenario. Natural disasters like hurricanes and earthquakes are likely to disrupt otherwise excellent communication, requiring a communication form that is beyond the interference of terrestrial or cellular systems (Holmstrom et al. 2011). In this context and in the range of communication possibilities, satellite communication offers an affordable approach to vehicle tracking, data transfer, as well as precise location services for the benefit of efficiency in a public safety communication.

13.4.4 Mobile Satellite Systems

Mobile satellite systems (MSSs) refer to communication systems delivered to and from mobile devices or locations (Elbert 2008). Typically, they include maritime, land, and aeronautical services (Basari et al. 2010). However, the current focus is mainly on a land mobile satellite service.

13.4.4.1 Vehicle-Mounted Mobile Satellite Communications Systems
According to Evans (1999), the conventional nature of mobile satellite communication is such that it was configured for application in the maritime context. Essentially, it is

Figure 13.3 Vehicle-mounted and flyaway satellite systems for search and rescue, *Ground Control Global Satellite Communications* (2018). Source: Retrieved from http://www.groundcontrol.com/prod_ig2500_001.htm.

communication among mobile use through satellite as opposed to cellular use (Evans 1999). The emergence of radios enabled the perpetuation of land based options for mobile satellite communication, facilitating the transmission of information in areas with disruptions to existing terrestrial systems (Evans et al. 2005). The advances have crossed into the efforts of public safety communication, specifically in the need for the delivery of emergency care in areas constantly lacking cellular or terrestrial communication. Consequently, one of the most dominant approaches features the mounting of the antenna on vehicles to facilitate safety communication (Figure 13.3).

The introduction of vehicular mountings for mobile satellite communication options is motivated by the range of communication services that may be required in otherwise impossible situations. Some areas require fast deployment, but the cellular and terrestrial set up may be hindered by natural conditions or disasters (Excelerate Technology 2018). Satellite broadband has the capacity to deliver the guarantee of communication in these areas, even in the absence of terrestrial communications, or where they may exist but they are congested or disrupted (Excelerate Technology 2018). Technically, the approach features mobile vehicles with the capacity to act as receptors and command centers. As these vehicles move about a given area, total and ongoing user support are possible despite the potential disruptions of events such as storms (Basari et al. 2010). Vehicle mounting of satellite communication systems promises the capacity to traverse rural areas with poor cellular reception for the maintenance of constant public communication (Urgent Communications 2010). This is especially due to the capacity to make audio transmissions even in situations of immensely low bandwidth, a feature that facilitates activities such as emergency dispatch and reports of progress (Evans et al. 2005).

Land MSS, therefore, apply a common principle toward their effective application. The service comprises of personal location beacons that act as the earth stations, often located on vehicles such as trucks, ambulances, or even trains (Basari et al. 2010). The land mobile satellite service facilitates a connection between the communication earth system and the vehicle system, ensuring the consequent transmission of data through the satellite. The configuration of the MSS remains constant, featuring a satellite, a fixed earth station, and a mobile earth station as part of the segments (Maral and Bousquet 2011). The channel quality, however, is advanced by the introduction of the propagation path that anticipates blocking by buildings or shadowing by foliage (Maral and Bousquet 2011). As such, the focus of this system is on ensuring that reception is direct as opposed to the use of the reflecting signals.

The specific configuration of the mobile station is also relevant, featuring an antenna, a diplexer (DIP), a high power amplifier, as well as a low noise amplifier (LNA) (Basari et al. 2010). The components also feature the up-converter and down-converter (U/C and D/C) alongside the modulator (MOD) and demodulator (DEM) (Basari et al. 2010). These features demonstrate the fact that the mobile earth station and the fixed one have similar configurations, and the presence of the MODEM can facilitate transmission without relying on the transponder.

Multiple innovations have focused on the provision of mobile satellite communication, especially through vehicular mounting. For instance, the Japan Aerospace Exploration Agency (JAXA) indulged in the development and launch of the largest geostationary S-band satellite (Basari et al. 2010). The satellite, named the Engineering Test Satellite-VIII (ETS-VIII), has proven effective in the transmission of messages, especially to rural areas and emergency services (Basari et al. 2010). These areas are unaddressed by the existing systems, in particular due to the typical lack of cell coverage in rural areas. Terrestrial systems are also likely to fail due to mechanical damage, especially in times of physical disaster (Desourdis et al. 2001). Several services are possible through the application of the vehicle-mounted satellite systems. In the event of a disaster, the use of vehicle-mounted systems such as the ETS-VII proves crucial for the communication of messages (Elbert 2008). Therefore, actions such as data communication, voice broadcasting, and message communication for disaster relief are executable through these satellite systems.

It is crucial to understand that the use of the vehicle-mounted satellite communication systems may prove to be a challenge from some dimensions. For instance, the use of large parabolic dishes on these vehicles may prove impractical, prompting the use of much smaller antennas (Iida 2003). Larger dishes have implications on the streamlined shape of the vehicle, which consequently influences aspects such as fuel consumption and efficiency. Therefore, the smaller antennas are often used on these vehicles in facilitating public safety communication (Basari et al. 2010). The implication is a reduction in the antenna gain, which puts limitations on the potentially receivable bandwidths. These issues could prove detrimental in the course of emergency operations, especially where the awareness of the most viable bandwidths is lacking. Nevertheless, alternatives such as larger vehicles that diminish the impact of large parabolic dishes on movement efficiency could offer partial solutions to the challenge (Iida 2003).

At the same time, additional challenges to the use of vehicle-mounted mobile satellite may feature the constancy in the change of direction. The vehicle-mounted antenna has to be omnidirectional to facilitate efficiency in the application of the technology specifically in the provision of emergency care (Basari et al. 2010). The alternative is the use of antennas with such tracking capabilities to maintain its boresight direction toward the satellite despite the changes in direction (Iida 2003). Contentious matters emanate from this perspective. The omnidirectional satellite has less gain than the parabolic dishes, severely restricting the signals received and applied in the communication effort (Maral and Bousquet 2011). At the same time, significant costs will be associated with the mounting of an antenna with azimuth rotation on top of a vehicle, which often leaves the option of either encountering the costs or risking the limitations to satellite reception and efficiency of communication (Iida 2003). The former may prove unsustainable depending on the budgetary implications of the public safety communication initiatives, while the latter may diminish efficiency in safety operations.

Mobile satellite communications were initially designed for application in maritime settings. Their efficiency thrives on the open sea and the absence of any obstructions except, perhaps, rain clouds (Urgent Communications 2010). However, unlike in the maritime context, the use of vehicle-mounted satellites in public safety communication faces the challenges of vehicles constantly moving behind barriers such as buildings and foliage (Iida 2003). These barriers could have detrimental effects on the efficiency of communication as well as messaging content. Consider, for instance, communication regarding the location of an injured party. The location changes between the first and second communication but the first communication was obstructed and only played out after the second was received (Iida 2003). The outcome would be the hindrance of service provision, threatening the welfare of the victims and compromising the effectiveness of safety communication efforts.

Therefore, despite the promising outcomes of this approach to using satellite communications in public safety, there are some significant challenges for consideration. Nevertheless, these challenges are not insurmountable, despite having extensive cost implications. In the presence of sufficient budgetary allocation, public safety providers can effectively apply the technology to overcome the shortcomings of cellular and terrestrial communication options.

13.4.4.2 Emergency Communications Trailers

Emergency communication trailers are one of the most common applications of mobile satellite communication systems. The concept features fully self-contained wireless terminals for communication, allowing the onboard generation of power through present equipment, as well as details including long-range Wi-Fi access points (Evans et al. 2005). There are also additional components such as ethernet switching routers alongside equipment to facilitate satellite RF transmission. The satellite communication trailers are designed to support low traffic mobile broadband services, often proving useful in areas with temporary blackouts in mobile coverage (Basari et al. 2010). Typical operations have been among miners and government military operations. Nevertheless, the flexibility and portability of the trailer alongside its transmission capabilities open it up to the application in public safety communication (ICS 2018).

Specific communication trailers have different capabilities, such as the product by Ground Control. The T-100 communications trailer is a self-contained multi-purpose vehicle, produced by the company Ground Control (Ground Control 2019). Its nature ensures its quick deployment to any location for the creation of high-speed (20 × 5 Mbps) internet connection through satellite (Ground Control 2019). This trailer is also applied toward the establishment of a half-mile wireless access point for any devices that are in range with wireless capabilities (Urgent Communications 2010). This communications equipment is capable of providing essential internet and voice over internet protocol (VoIP) phone service five minutes after parking, allowing the release of the towing vehicle until the required time for more movement (Ground Control 2019).

13.4.4.3 Flyaway Satellite Internet Systems

Flyaway satellite internet systems are a class of small and portable satellite uplink terminal systems applied toward the broadcast of communication in isolated locations (Rutenbeck 2012). As self-contained uplink alternatives, they essentially apply the same technology as well as the choice of equipment such as vehicle-mounted automatic pointing mobile satellite units (Mobil Satellite Technologies 2018). However, they are different on the basis of the deliberate portability introduced into the systems through installation onto shippable and waterproof ruggedized cases (Mobil Satellite Technologies 2018). The design of these units is such that they can be transported over great distances or any form of terrain without damage, immediately responding to the needs for broadband connectivity deployment within moments (Excelerate Technology 2018). The reinforced, so to speak, nature of the systems implies the presence of all the benefits of the vehicle-mounted technology with fewer risks of destruction. The implication is especially significant here when addressing public communication in the context of storms. The ruggedized nature ensures that the systems remain functional despite exposure to conditions of physical impact or even rain, subsequently maintaining service provision despite otherwise challenging circumstances.

The application of the flyaway satellite internet antennas in public safety communication depends on the engineering capacities that serve as the improvement modifications (Urgent Communications 2010). The details reveal the capacity for high engineering on these systems, featuring aspects such as multi-part carbon fiber segmented reflectors and feed arm assemblies that are removable (Excelerate Technology 2018). These aspects make them not only capable of use in high-pressure scenarios, but it also eases the process of transportation from one location to the next. Additional engineering modifications for the systems include quick disconnect RF connections as well as removable block upconverters (BUCs), all features that combine to make the systems easy to move and use (Mobil Satellite Technologies 2018). Despite their nature in forms that resemble the vehicle-mounted systems, flyaway satellites allow mounting on the ground. This element implies relative ease in set-up, with the provision for setting up not only on a roof rack but at any point on the body of the vehicle.

Like the vehicle-mounted MSS, the flyaway internet systems provide the technical capacities to allow operations by an individual. The systems can be applied toward the deployment of emergency services such as ambulance location and direction, as well as the control of dispatch in operations such as fire control. At the same time, the capacities for the systems to transmit both analog and digital information imply the capacity for multiple forms of information sharing. These forms include the sending of fax and email as well as the sharing of video and audio content via satellite transmission (Mobil Satellite Technologies 2018). In situations of constant communication and transmission, the use of these systems introduces not only convenience and portability but also expands the capacity for control of operations by a single individual. These advantages are crucial especially in contexts when the budgetary implications of the venture into public safety communication may be negative.

The capabilities of these systems have been tested in the commercial and military contexts. They report reliability for a wide range of operations or SATCOM applications (Basari et al. 2010). The application of the terminals in the modern battlefield communications context has also proven beneficial. These premises have formed the foundation for the inclination of production alternatives toward systems that respond specifically to the context of public safety.

13.4.5 VoIP Phone Service Over Satellite

The VoIP service is applicable to the provision of phone voice services through satellite. The use of VoIP over satellite is based on the typical processes of transmission that characterize satellite internet systems (Figure 13.4). There are beliefs that the VoIP service does not work over satellite but assertions from the professional perspective refute this claim as simply untrue (Miller 2017). Tests have also confirmed that satellite links have the capacity to carry VoIP, despite communication breaking down under some errored link conditions (Kota et al. 2011). Multiple service providers offer VoIP alternatives, even where they acknowledge the possible presence of shortcomings associated with the provision of these satellite services. These shortcomings are inclusive of delays in the ensuing communications, induced by the latency emanating from the nature of satellite communication (Elbert 2008).

While there are efficient communications taking place through cellular services, the use of VoIP is crucial as an alternative for areas with poor cell reception. The execution of communication in areas with poor cellular reception relies on the use of landline for voice calls, which could be the terrestrial version or the satellite version of phone service (Miller 2017). Therefore, as long as the satellite signal bean is available to a particular user, they are able to make voice calls or use internet services like Skype for

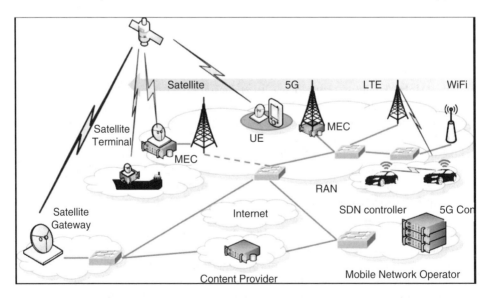

Figure 13.4 View of a satellite internet system (Höyhtyä et al. 2017).

communication. Evidently, this alternative form of communication may prove beneficial in areas where cellular reception is inhibited but the need for safe operations and communications persists (Miller 2017). The coordination of police work or the fire department in rural areas, for instance, could benefit from this form of service provision. Calls may take place over this satellite network, and information sharing could also take place through capabilities like Skype. The latter is particularly beneficial in cases such as guided rescue operations where direct vision for the dispatcher is a contributing aspect of the effectiveness of their efforts (Kota et al. 2011).

Satellite links have increasingly become an integral part of global communications, especially in enabling reach in remote areas. Advances in satellite systems setup have enabled overcoming the shortcomings of satellite VoIP (Elbert 2008). There are alternatives such as dedicated backbone channels that diminish occasions of network intolerance for radio usage. Direct way users also have alternatives such as the perfect bandwidth to enable consistent communication (Kota et al. 2011). As new satellite systems continue being introduced for the provision of internet access for home users and businesses, the inclusion of VoIP has become a near constancy in the new innovations. The performance of VoIP under satellite, nevertheless, will often vary under various loading conditions.

There are some challenges that may accompany the application of VoIP for communication in the maintenance of public safety. These are, primarily, challenges relating to interoperability, especially in commercial options, where various gateways challenge the possibility for interoperable voice transmissions (Miller 2017). At the same time, IP networks may often lack the capacity for VoIP services at levels similar to the same service in the public switched telephone network (PSTN) settings (Kota et al. 2011). However, where these challenges are overcome, it is possible to apply the VoIP satellite services for call center integration as well as unified messaging. The main control, therefore, can use a single link to acquire information of aspects such as disaster management while still continuing matters such as dispatch (Maral and Bousquet 2011). Similarly, it is possible to apply unified messaging in cases of multi-agency projects such as disaster relief efforts. These outcomes, nevertheless, are only achieved in the presence of deliberate efforts to integrate the service and achieve sufficient interoperability among the devices or systems.

13.4.6   Fixed Satellite

The fixed satellite service is among the oldest and most common of the existent satellite services globally. The service is intended for communication between fixed earth stations using a satellite, or specifically for handling communication within a limited area. Understanding the concept of fixed satellite services is crucial, as they refer to earth stations that are not in motion while in use as opposed to earth stations that are permanently immoveable (Chartrand 2004). This is in contrast to mobile satellite services, such as in vehicle-mounted forms, whereby usage takes place as the satellite remains in motion. The satellite forms are exemplified by corporate networks, which address communication within a limited area, as well as very small aperture terminals (VSAT), and distribution networks (Leonard 2010). Satellite newsgathering exemplifies the occasions when the fixed satellite services can actually be moved, creating a new location for the fixed-satellite service communication, albeit with a shift in the location of the earth stations.

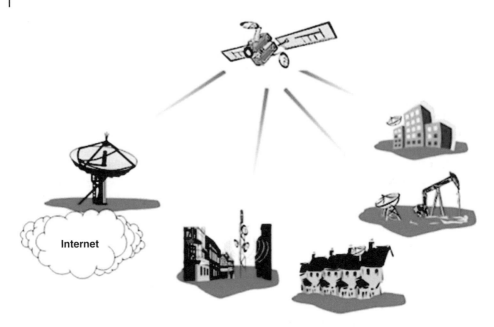

Figure 13.5 Fixed satellite communications. Source: ASDNews (2015).

The operation of the FSS frequency features the specification of the direction of the travel signal. One of the bands handles the uplinks – Earth to space – while the other handles the downlinks, referring to space to Earth (Kota et al. 2011). Notably, these frequencies have to be different in order to prevent their interference with each other. Despite the difference, the bandwidth remains the same to facilitate the equal transmission of information. Compared to the mobile satellite services, the fixed satellite handles larger bandwidth due to the expected roles it executes such as television content (Leonard 2010). As shown in Figure 13.5, the FSS enables communication between earth stations at given positions. It is possible to have a single specified point or a specified area of communication that allows transmission within the range of the point. The network topology, therefore, ensures that communication can take place either locally or even at national and international levels (Kota et al. 2011). Configurations of FSS, therefore, can be crucial to the establishment of a national framework for the coordination of public safety communication efforts.

In the context of public safety communication, relying on the fixed satellite services would be crucial to the transmission of data signals or highly multiplexed telephony. The typical terminal size features a 1.2 m dish with a modem and RF equipment, translating to weights above 100 lb (Chartrand 2004). The application of this service is effective at command center communications for the continuity of operations, as well as any of the activities that would require the speedy exchange of information. When the FSS is applied as a shared service, it may prove more economical than the effort by individual agencies to apply their own satellite transmissions (Kota et al. 2011). Developments in FSS service are predicted to expand both in applicability and market size. The global market for fixed satellite services is projected to reach US$24.02 billion by 2025, perpetuated by the increase in both public and private services (Department of Homeland Security 2014).

The use of fixed satellite service communication is recommended for continuity and crisis management as well as for the purposes of homeland security. The current and future adaptability of satellite communication services is expected to become an integral part of designing and implementing safety communications, despite advances into other communication alternatives (Kota et al. 2011). Fixed satellite communications can prove especially crucial to the first hours of disaster management, especially where conventional communication has been disrupted, specifically the terrestrial alternatives. Several public health services exploit the FSS communication both at fixed locations as well as on moving vehicles (Maral and Bousquet 2011). The framework offers a sustainable and cheap approach to the maintenance of public safety communication and emergency preparedness on a consistent basis.

13.4.7 Frequency Allocations in FSS and MSS Systems

Multiple satellite communication services have different frequency allocations. The allocation of frequency bands is toward the achievement of compatible use. Due to regional differences, the allocations may be shared or overlap and often exhibit different degrees of saturation (Maral and Bousquet 2011). Several of the bands are also reserved for specific administrative purposes like government while others are useful in operational contexts like the pursuit of public safety groundwork. The FSS typically has frequency bands including 6 GHz for the uplink and 4 GHz for the downlink (C band) (Elbert 2008). These frequencies are occupied by older systems such as domestic satellite communications and INTELSAT (Elbert 2008). It is possible that these systems are either close to obsolete or excessively saturated for effective public safety communication use. Other frequency allocations for the FSS include the 8/7 GHz (X band), 14/12 GHz (Ku band), 30/20 GHz (Ka band), and the 50/40 GHz (V band) (Chartrand 2004). The Ku band reflects conventional operational applications. Nevertheless, there is escalating interest in technologies supporting frequencies above 30 GHz due to minimal interference and their capacity to deliver quality messages (Maral and Bousquet 2011).

Frequency allocations also differ in the context of mobile satellite services. The L band under MSS has allocations of 1.6/1.5 GHz, applied in geostationary systems (Maral and Bousquet 2011). The Ka band under the FSS is also transferable to the context of mobile satellite services. Under the broadcasting satellite services (BSS), frequency allocation features 2/2.2 GHz under the S band – crucial to the application of mobile communications internationally (Maral and Bousquet 2011). The Ku band under MSS operates with the 12 GHz frequency allocation. While these allocations are typical of the geostationary systems, there are allocations for non-geostationary systems. The allocations include very high frequencies (VHF) and ultra-high frequency (UHF), with bandwidths spanning 138/150 MHz for the former and 401/460 MHz for the latter (Maral and Bousquet 2011). Therefore, the decisions toward the specific use of either FSS or MSS in public safety communications will rely on the available architecture as well as the implications of the bandwidth for the quality of communication.

There is evidence of increased technological capacities among these alternatives. With changes in conventional communication from analog to digital, the capacity to achieve flexible connectivity between beams presents as a crucial improvement to the shared communications. The application of higher bands has also facilitated the provision of broadband services. However, these services may encounter challenges such as interference in the case of rain (Richharia 2017). As progression in technological developments continues, the harnessing of even higher bandwidths promises more efficiency in the application of the satellite services in the context of public safety communication.

13.5 Limitations of Satellite for Public Safety

While satellite communication options undoubtedly present solutions for the current and future contexts of public safety, there are some challenges that this alternative provides. One of the most prominent issues is the occurrence of propagation delays. Configurations of satellite communication in the star network usually report a total delay between two VSAT units by about 0.5 s, especially where communication takes place through the hub (Singh 2019). The delay is about 0.25 s in the case of communication directly between one VSAT and another. The delays are especially inconvenient in voice communications, especially as the result is the occurrence of voice overlap as well as echoes and possible misunderstandings of the message during emergencies (Richharia 2017). Data communication protocols such as the TCP/IP may also be negatively affected by propagation delays (Singh 2019). Therefore, the use of satellite networks usually demands the design of special protocols for effective data communication.

At the same time, there are limitations of low bandwidth support among the satellite options. Both mobile and fixed satellite options report lower bandwidth capacities than terrestrial options like optic fiber (Singh 2019). Conventional satellite communications have limited bandwidth capabilities of about 75 MHz, while fiber optics report capabilities that could be termed as unlimited at 50 THz (Richharia 2017). The implication is that communications using fiber optics rarely report interference while satellite communication may frequently encounter interference. The limitations could limit the efficiency of communication, both in the speeds of data information transmission as well as in the clarity of voice and video messages in the application of satellite for public safety (Singh 2019). This could suggest poor coordination for the teams in the field especially when emergency service provision covers a wide area.

There are limitations associated specifically with the use of LEO or MEO versions. The visibility from earth tends to be extremely limited, indicating the need for fast handover for effectiveness (Wireless World 2012). The outcome is the need for a large number of satellites covering the entire radius of the Earth. Ultimately, the system becomes extremely complex and expensive to maintain. Similarly, all satellites have a limited life of about 12–15 years (Wireless World 2012). The limitation creates the need for the preparation of another launch before the first satellite fails. A large number of satellites poses complexities as well as expensive implications. Setting up the system may prove inherently expensive, which could be compromised in case the public safety communication budgetary allocations are limited. For instance, national communication systems that rely on satellite communications to coordinate activities of the Department of Homeland Security (DHS) may face budgetary constraints if funding to the DHS is reduced. The dimensions could pose the risk of satellites becoming obsolete or the complex system being compromised, threatening handover and consequently limiting the efficiency of communication.

It is, however, notable that issues concerning quality and reliability are resolvable using engineering improvements, protocol gateways provide good services to the user, especially through the standardizing of the satellite protocol through the radio link. In the absence of the gateway, however, the standard protocol may fail to be communicated to the end user computer. Increased bandwidths to satellite links also have the capacity to overcome most of the performance limitations associated with satellite communication. Therefore, in a context where satellite communication is sufficiently funded for public safety, it is possible to make engineering design improvements that will effectively transform the performance of individual and shared inter-agency operations.

13.6 Conclusion

Public safety agencies continue recognizing the viability of satellite communication, an option once considered too pricey and exotic for functional capacities. Several reasons have perpetuated this inclination toward the consideration of satellite communication as a viable alternative. Primarily, advances in technological development have facilitated the decline in size and cost, while allowing greater capabilities in the transmission of voice and data. Consequently, the cost that public agencies would experience diminishes significantly as the communication equipment becomes more readily available.

Communication in public safety is evidently an area of profound concern. Agencies globally are seeking solutions in this dimension, exemplified by the National Public Safety Broadband Network (NPSBN) initiative. However, the solutions are still confounded by the shortcomings of terrestrial systems. While the world does seem to be gravitating toward Long Term Evolution (LTE) as the solution for public safety communication, the limitations imposed by low network coverage outline the opportunities for the use of satellite communications. Therefore, the current framework demonstrates the potential for the transformation of satellite communication into the context of public safety. GEOs and LEOs offer different capacities in the forms of speed, cost, and prevention of communication interference. Similarly, it is possible to implement the fixed or mobile satellite architectures, depending on the desired coverage and coverage content. Specific commercial options have maximized the approaches to portable satellite communication such as through vehicle-mounted and flyaway satellite, especially for the provision of internet capabilities. Consequently, these multiple forms give potential for continued communication even when existent terrestrial and cellular communication is hindered by weather or physical challenges.

Nevertheless, objectively reviewing the conditions for the use of satellite in public safety communication shows some significant barriers. Satellite networks are still potentially more expensive to implement, especially due to the need for constant maintenance and obsolesce in the long term. Communication quality may also be hindered by bandwidth limitations, delays, as well as issues in satellite handover. However, with a focus on incorporating these technologies into the context of public safety, these limitations are possible to overcome. Engineering improvements such as protocol gateways can ease the issues of delays in TCP/IP transmissions while the use of mega-LEOs offers one solution for bandwidth problems among LEO satellites. Commitment to the funding of the systems also will facilitate overcoming the conventional challenges of using this form of transmission. Therefore, in a time when public safety communication continues gaining prominence, the place of satellite communications and possibilities like vehicle tracking and emergency communication is evident.

References

ASDNews (Aerospace and Defense News) (2015): The Key Players in Fixed Satellite Service Market in North America 2015–2019

Baldini, G., Karanasios, S., Allen, D. et al. (2014). Survey of wireless communication technologies for public safety. *IEEE Communications Surveys and Tutorials* 16 (2): 619–641.

Basari, B., Saito, K., Takahashi, M. et al. (2010). Antenna system for land mobile satellite communications. *Satellite Communications*: 33–59.

Carl, L., Fantacci, R., Gei, F. et al. (2016). LTE enhancements for public safety and security communications to support group multimedia communications. *IEEE Network* 30 (1): 80–85.

Chartrand, M.R. (2004). *Satellite Communications for the Nonspecialist*. London: SPIE Press.

Desourdis, R.I. Jr., Dew, R.A., O'Brien, M., and Hinsch, H. (2015). *Building the FirstNet Public Safety Broadband Network*. Norwood: Artech House.

Desourdis, R.I., Smith, D.R., and Speights, W.D. (2001). *Emerging Public Safety Wireless Communication Systems*. Washington: Artech House.

Elbert, B.R. (2008). *Introduction to Satellite Communication*. Washington: Artech House.

Evans, B.G. (1999). *Satellite Communication Systems*. Washington: IET.

Evans, B. et al. (2005). Integration of satellite and terrestrial systems in future multimedia communications. *IEEE Wireless Communications* 12 (5): 72–80.

Excelerate Technology Ltd. Vehicle-Mounted Satellite Communication. 2018. 17 March 2019, St Mellons, UK: https://www.excelerate-group.com/products/vehicle-mounted-satellite-communications.

Ferrús, R. and Sallent, O. (2015). *Mobile Broadband Communications for Public Safety: The Road Ahead Through LTE Technology*. New York: Wiley.

Ground Control. (2019.) Public Safety - Emergency Communications via Satellite. 1 March 2019, San Luis Obispo, CA: Ground Control, http://www.groundcontrol.com/Mobile_Command_Communication.htm.

Holmstrom, J., Rajamaki, J., and Hult, T. (2011). The future solutions and technologies of public safety communications-DSiP traffic engineering solution for secure multichannel communication. *International Journal of Communications* 3: 115–122.

Department of Homeland Security (2014). *FirstNet: Building a Nationwide Public Safety Broadband Network TechNote*. Washington, DC: Department of Homeland Security.

Marko Höyhtyä, Tiia Ojanperä, Jukka Mäkelä, Pertti Järvensivu, Sami Ruponen. Integrated 5G Satellite-Terrestrial Systems: Use Cases for Road Safety and Autonomous Ships, 2017, 23rd Ka and Broadband Communications Conference, Trieste, Italy, October, 2017.

Hussaini, Umair. Kepler's Laws for Satellites – Space Segment. 2017. Technobyte https://www.technobyte.org/satellite-communication/keplers-laws-for-satellites-space-segment.

ICS (2018). *Satellite Trailer*. Thomastown VIC: ICS Industries www.icsindustries.com.au/products/communication-trailers/satellite-trailers.

Iida, T. (2003). *Progress in Astronautics and Aeronautics: Satellite Communications in the 21st Century: Trends and Technologies*. Washington: AIAA.

Kota, S.L., Pahlavan, K., and Leppänen, P.A. (2011). *Broadband Satellite Communications for Internet Access*. New York: Springer Science & Business Media.

Kumbhar, A. et al. (2016). A survey on the legacy and emerging technologies for public safety communications. *IEEE Communications Surveys and Tutorials* 19 (1): 97–124.

Latifi, R. (2011). *Telemedicine for Trauma, Emergencies, and Disaster Management*. Miami: Artech House.

Leonard, B. (2010). *Connecting America: The National Broadband Plan*. DIANE Publishing.

Maral, G. and Bousquet, M. (2011). *Satellite Communications Systems: Systems, Techniques, and Technology*. London: Wiley.

Miller, Alex. Does VoIP phone service work with satellite internet? 2017. Carlsbad, CA: Viasat https://www.exede.com/blog/voip-phone-service-work-satellite-internet.

Mobil Satellite Technologies. (2018) Flyaway Satellite Internet. Chesapeake, VA: Mobil Satellite Technologies. https://www.mobilsat.com/Flyaway-transportable-satellite-internet.

Richharia, M. (2017). *Satellite Communication Systems: Design Principles*. New York: Macmillan International Higher Education.

Rutenbeck, J. (2012). *Tech Terms: What Every Telecommunications and Digital Media Professional Should Know*. Chicago: CRC Press.

Singh, Soneshwar. Addressing the Big Picture Challenges of Satellite Communication. 2019. Hackernoon https://hackernoon.com/addressing-the-challenges-of-satellite-communication-40ded030d9b1.

Urgent Communications. Satellite communications for public safety. 2010 IWCE's Urgent Communications https://urgentcomm.com/2000/01/01/satellite-communications-for-public-safety.

RF Wireless World. Advantages and disadvantages of Satellite Communication. 2012. RF Wireless World http://www.rfwireless-world.com/Terminology/Advantages-and-Disadvantages-of-satellite-communication.html.

(2018). Ground Control Global Satellite Communications. In: *Mobile Satellite Internet & Phone*. http://www.groundcontrol.com/prod_ig2500_001.htm.

14

Public Safety Communications Evolution: The Long Term Transition Toward a Desired Converged Future

14.1 Introduction

14.1.1 Toward Moving Public Safety Networks

Fourth generation Long Term Evolution (LTE) stands to be the technology selected by the United States federal and European Union authorities to be used by public safety networks to assist during various emergency scenarios. Since the beginning of Release 11, the Third Generation Partnership Project (3GPP) has undergone massive developments to enable scalable, robust, and resilient nationwide public safety broadband networks that are crucial during emergency services. The chapter will discuss and highlight the possible public safety use cases with the existing network topologies, the nature of 3GPP standards, and future challenges.

In the United States as well as other countries within the European Union, several vendors such as Ericsson, Nokia-Alcatel, Huawei, Cisco, Motorola, and Thales are now championing for the use of LTE-based public safety solutions. Because the current technologies are moving to higher bandwidth, the existing public safety solutions such as Project 25 and terrestrial trunked radio (TETRA) cannot meet the demands and requirements of providing effective and efficient mission-critical voice communicators because of their pre-existing design (Sharp et al. 2004). Additionally, the original design for the LTE system, being a commercial cellular network, was not suited in the first 3GPP releases. This is in line with the support of public safety services such as reliability, confidentiality, security, and group and device-to-device communications. Based on this, the gap created by the discussion, or rather the most important question, is whether LTE is an appropriate network to support public safety. Therefore, to address these issues, several studies on device-to-device communications, mission-critical push-to-talk (MCPTT), as well as the isolated evolved terrestrial radio access networks (E-UTRAN) have been released (Sharp 2014).

14.1.2 The Communication Needs of Public Safety Authorities

As stated by Sharp (2014) the communication needs of public safety authorities include

➢ Reliable and robust voice communication everywhere
➢ Allowing for co-operation and easy communication among all enterprises
➢ Short messaging for alarming, field task delivery and securing the validity of information

Public Safety Networks from LTE to 5G, First Edition. Abdulrahman Yarali.
© 2020 John Wiley & Sons Ltd. Published 2020 by John Wiley & Sons Ltd.

> File transfer from the place of incident to support sites as the command or 112 canters
> Establishing a platform for communications from the field for daily office work.

The majority of urban setups require communication to ensure effective and efficient services during emergencies. It is the responsibility of the government to ensure that citizens are safe and can readily secure emergency services. This can only be achieved if public safety agencies are efficient and reliable. It is important to note that professionals are heavily reliant on mission-critical communications to assist in working together through engaging in situational awareness response times and control when faced with a challenging situation. A majority of taskforces require reliable and efficient communication systems, especially during emergency scenarios.

It is vital for information to be shared via radio in a timely and reliable way. This is because many operations within the public safety agencies such as fire departments and law enforcement are heavily dependent on them. Therefore, public safety wireless networks (PSWNs) active within the scenes of emergencies play an important role. According to Rajaratnam and Takawira (2001), the public safety community has established certain factors that limit the efficiency and effectiveness of public safety networks. One of the main issues of public safety includes the limitation of the fragmented radio spectrum. For instance, in the United States, the establishment of the SAFECOM project was to increase the efficiency and effectiveness of wireless communications within the local, state, and federal public safety agencies (SAFECOM). Meanwhile, it is a fact that the emergency wireless communication service providers are often concerned with factors such as spectrum shortage and high deployment costs. Taking an example of Vancouver, the emergency agencies in the area only share wireless voice channels (Nikaein et al. 2014). Hence, the importance of this topic is to help in understanding public safety communications through analyzing, evaluating, and optimizing the use of these networks toward a converged future.

14.1.3 The Nationwide Public Safety Broadband Networks

The concept of a public safety broadband network (PSBN) refers to the communication applications of secure high-speed wireless data communications within a single network (Desourdis et al., 2015). Its application in emergency contexts is only part of the viable alternative, presenting opportunities for communications in emergencies as well as the daily operations surrounding the choice agencies. The nature of this type of network is such that it has the potential for the improvement of the effectiveness of first responders as well as their safety and that of the public (Ferrús and Sallent 2015).

As such, the Nationwide Public Safety Broadband Network (NPSBN) initiative is the first of its kind, seeking the building of a cellular network that prioritizes application in public safety. Its conception in 2012 was from the enactment of the Public Law 112-96-Middle-Class Tax Relief and Job Creation Act, part of which featured this approach to emergency communication (Ferrús and Sallent 2015). Its success is not only dependent on the actions of the entity in charge, FirstNet, but also on the reliability of established rural broadband coverage and implications for public safety. FirstNet is an independent authority within the US Department of Commerce in the department of telecommunications (Desourdis et al, 2015). As such, all actions necessary toward the design and implementation of the national network along with the consultations for collaboration remain the responsibilities of this entity. While all public safety organizations or agencies have access to this network, it cannot be leased to another entity for secondary use (Ferrús and Sallent 2015).

Perhaps the most pertinent detail surrounding this network is the choice of architectural features, especially considering the implications of various forms of communication on interoperability. The NPSBN mainly applies LTE technology, occupying Band Class 14, which is the public safety section of the 700 MHz wireless spectra (Department of Homeland Security, 2015). There are multiple components characteristics of this communication network including the presence of a core network; radio access networks (RANs), as well as transport backhauls (Desourdis et al., 2015). There is also the extensive application of public safety devices under the spectrum of this new broadband network.

Specific functions relate to the elements of the overall architecture. Primarily, the core network processes, stores, and ensures data security (Department of Homeland Security 2015). It is also in charge of activities such as the hosting of specific applications and services while also facilitating the interfacing with other networks at the local, state, and federal levels inclusive of the 911 emergency systems (Desourdis et al., 2015). The transport backhauls connect the core to the RANs, featuring the use of fiber optics or coaxial cables. Specifically, the RAN infrastructure features mobile hotspots and cell towers expected to facilitate connection to public safety devices incorporated into the system including tablets, laptops, and smartphones (Desourdis et al., 2013).

It is notable that the introduction of the NPSBN, while seemingly making use of conventional devices, may require the introduction of new devices for compatibility. This is indicative of expenditure in the billions in an effort to ensure comprehensiveness and overcome the challenges of interoperability facing the use of the land mobile radio (LMR) in isolation. The devices require compatibility with the Band Class 14 spectrum, which implies a responsibility to manufacturers for production to address this market niche (Department of Homeland Security 2015). Evidently, there are implications for production where manufacturers may be reluctant to engage in the process, considering the relatively small market base. There is an alternative approach involving the incorporation of the capabilities within the existent smartphone and laptop technology (Department of Homeland Security 2015). However, the latter approach implies exposure to security threats such as hacking and denial of service in times of emergencies.

While the system is broadband based, there are challenges for the use of LTE technology, especially in the coverage of rural areas. The proximity of cellular towers in urban areas also significantly compromises communication in some contexts (Department of Homeland Security 2015). It is on this basis that the application of satellite communication technology gains prominence as an alternative to the possible interruption to LTE communication. Alongside options like terrestrial communication systems, satellite communication will ensure coverage in all areas and facilitate efficiency in the process of emergency management (Desourdis et al., 2015). However, the effort also involves the deliberate hardening of its infrastructure to ensure resilience to conditions like earthquakes and storms in the different areas within the US. This is alongside redundancy initiatives like backup sources of power as well as diversification in the routing of the networks (Ferrús and Sallent 2015).

As such, the pursuit of this network is a significant step toward the achievement of convergence of voice and data in mission-critical communications. The approach allows the maintenance of LMR networks while introducing broadband alternatives to support image and data transmission (Department of Homeland Security 2015). With indications of the continued evolution of communication efforts in the public emergency context, it is evident that the solution for a dedicated service is only viable through the transition into the public safety network services (Ferrús and Sallent 2015). Deployment may take time, especially due to the need to harness solutions like satellite backup for network routing, which is the basis for the continued use of the LMR communication in safety settings until convergence is achieved.

14.1.4 Global Public Safety Community Aligning Behind LTE

The engagement of the USA in the implementation of the NPSBN conceptualizes one of the global efforts toward the use of LTE in the management of public safety communications. Indications are that, globally, the public safety market using LTE is likely to remain lucrative with an annual growth rate of about 22.07% in the years between 2018 and 2025 (Weny News 2019). Figure 14.1 shows a potential scenario based on LTE deployment for a public safety network. Governments have indicated growing interest in the use of wireless technologies for public safety communication, with the use of LTE technology beating any upgrades to existent 2G networks (Sharp 2014).

The outcome has been a perceptible consensus in the adoption of LTE frameworks on a commercial basis for public safety and security, allowing the possibility of a common standard and the pursuit of interoperability (Carla et al. 2016).

The development of public safety networks in the USA based on LTE has been alongside efforts at a global scale to pursue similar developments. These developments are alongside the adherence to particular safety standards. Primarily, TETRA is the leading standard for public safety beyond North America (Sharp 2014). The standard supports medium speed data but with the consciousness that advances in technology are essential to the addition of broadband technologies. The intention of TETRA, therefore, includes the addition of the LTE standard through the inclusion of required functionalities in

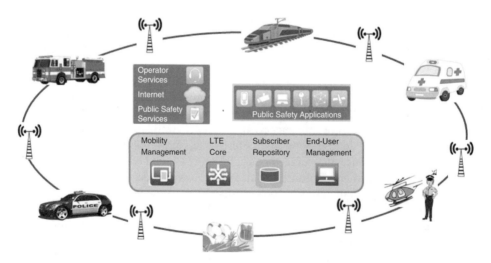

Figure 14.1 Potential for LTE in the global framework of public safety, retrieved from https://blog .technavio.com/blog/global-public-safety-lte-market-bound-for-success.

order to meet the objective of driving mobile broadband solutions for mission-critical communications (Liebhart et al. 2015).

Similar views regarding the use of LTE technology in public service communications manifest in the Tetra and Critical Communications Association (TCCA) and the European Telecommunications Standards Institute (ETSI) (Sharp 2014). ETSI especially emphasizes the use of LTE under standardization by 3GPP, allowing the use of both public and private networks in the process (Liebhart et al. 2015). Nevertheless, the adoption of LTE to aid the current TETRA standards is required to consider factors of interoperability, with approaches to ensure interoperable interactions between satellite communications and terrestrial segments (Liebhart et al. 2015). The support of these organizations for this communication network has functioned toward placing LTE as the next baseline technology for advances in public safety communication (Sharp 2014).

From a technical perspective, LTE standard enhancements to ensure their incorporation into areas of public safety are taking place under the 3GPP collaboration. Nevertheless, the national or local communication networks involved in public safety communication make significant contributions, especially in matters of interoperability (Liebhart et al. 2015). Therefore, the focus of this improvement is on the maintenance of the best LTE features while also ensuring its inclusion of features crucial to the processes of public safety communication. Therefore, specific features are required within the public safety framework: proximity services for interoperability and group call capabilities for dispatcher functions (Sharp 2014). With these features, however, it remains essential to include features that prevent malicious attacks on the system.

As such, an LTE framework for public safety and security promises the benefits of open security standards as well as a large number of organizations involved in security monitoring for the system. The chances of detecting hacks or backdoor increases with the presence of many countries involved in monitoring, reducing the risk of external interference (Liebhart et al. 2015). The LTE also introduces flexibility toward more open network sharing, which should facilitate the collaboration of public safety efforts in multiple countries (Liebhart et al. 2015). The application of RANs, in particular, enhances this network flexibility. Such an approach has advantages, especially in European countries where public safety communication may be designed per country but emergencies may require inter-country collaboration (Sharp 2014). Nevertheless, this latter component remains a subject of debate regarding the introduction of baked-in solutions for easy maintenance or the use of more customized features to match the specific emergency needs of individual nations (Sharp 2014).

14.1.5 Understanding the Concept of E-Comm in Relation to Public Safety

The history of E-Comm can be understood in relationship to the event of 1994 of the Stanley Cup riots in Vancouver. As in any emergency situations, several agencies in Vancouver town such as the police, fire departments, the Royal Canadian Mounted Police, and British Columbia Ambulance mobilized to deal with the emergency situations. However, after the situation was settled, it was established that the communication between these agencies was inadequate and not so reliable. For example, one of the main reasons that contributed to this was that the various agencies had their own wireless radio networks that were not interconnected and therefore interoperable. On the

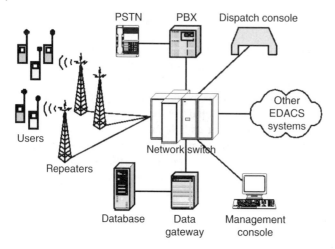

Figure 14.2 Systems architecture of EDACS (Song and Trajkovic 2005).

other hand, it was discovered that the emergency radio channels used by the police and the fire departments required a massive upgrade. As a way of resolving this problem, the British Columbia government started an emergency communication project, which later evolved into the E-Comm in 1997 (Huang et al. 2012).

Ideally, E-Comm is a regional emergency communications center for Southwest British Columbia, which mainly supports a wide area radio dispatching services during emergency situations. While looking at it from the public safety networks perspective, the E-Comm's wireless radio network uses enhanced digital access communications systems (EDACS) (Frost and Melamed 1994). Figure 14.2 shows the system infrastructure of EDACS. It is manufactured by the M/A Com and below is its architecture. The architecture has a central system controller or a network switch, a management console, many repeater sites or base stations (BSs), one or more fixed user sites (dispatch consoles), a private branch exchange (PBX) gateway to the public switched telephone network (PSTN) and several mobile users. It is also important to note that the individual systems or the cells within the E-Comm networks are interconnected by high-speed data links. All the system events and the call activities within the network are often recorded at the BSs at every hour through the data gateway.

14.2 Transmission Trunking and Message Trunking

Since it is a form of the public safety network, the nature of communication within the E-Comm takes place by transmission trunking, rather than the conventional method of message trunking. When it comes to message trunking, a radio channel is often given a call for the entire period the call takes place. However, it is crucial to note that there gaps in silent periods exist during conversations. Hence, in terms of channel utilization, occupying the entire channel during a conversation is inefficient. Whereas, in transmission trunking, the radio channel is automatically released as soon as the caller finishes the last sentence and releases the push-to-talk (PTT) button. This allows the next caller to start a call by pressing the PTT button in their radio device. Availability of the channel is then checked by the system and assigns the needed number of channels to the call. According to Sharp et al. (2004), transmission trunking has been found to be approximately 20–25% efficient as compared to message trunking. However, it is important to note that the high channel utilization comes with a higher price when using the transmission trunking mode. The overheads resulting from the channel assigning time and channel dropping time are added to each transmission. This is because the process of channel assigning and channel dropping are repeated any time the PTT button is pressed. The good thing is that the continuously available high-speed control channel within the E-Comm network handles this problem through alleviation. For instance, the control channel utilizes a dedicated radio channel that supports 9.6 kbps digital signaling, and the channel assigning and dropping times are 0.25 and 0.16 s respectively.

14.2.1 Push-to-Talk Mechanisms

The main communication mechanism in the E-COMM network is PTT. This means that any user who wishes to start a phone call needs to press a button on the radio device to request one or more channels. The work of the control switch is to identify users in the specific talk group. In the case that free channels exist, the caller will receive a credible signal so as to establish the call. Otherwise, it means that the call request will be queued (Casoni et al. 2015).

14.2.2 Talk Groups and Group Calls

Within the E-Comm network, talk groups are often categorized into various levels known as agency levels, fleet levels, and sub-fleet levels to help in enhancing the coordination of operations. Normally, each radio device belongs to one or more talk groups and may be switched between talk groups. The standard cell type in the E-Comm network is the group call. Call recipients are often members of a talk group. It is important to note that one of the main advantages of a group call is that it eliminates the need for radio users to know the target device number in order to reach an individual user. It is often common for the users to call a target group without having to know the existing and the current members of the group.

Consequently, it is significant to note that within the E-Comm network, a single call might use several channels simultaneously. This means that if all users of a talk group live within a single system, the network controller will allocate one free channel to the call. However, if members of a talk group are distributed across several systems, the network controller will have to allocate the call to a free channel in each system.

14.2.3 Mobility of Radio Devices and Call Handover

The two major concerns for micro-cell or cellular networks are mobility of radio devices and call handover. They do not affect the E-Comm system that much as they are of little importance.

14.2.4 WarnSim: Learning About a Simulator for PSWN

It is known that packet-switched networks require various simulation tools for simulation such as OPNET, ns-2, and J-Sim. However, it should be noted that these tools might

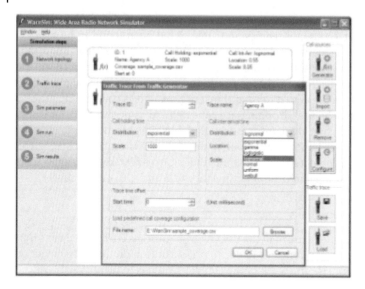

Figure 14.3 WarnSim traffic trace generation.

not prove effective while working with circuit-switched networks such as PSWNs. To deal with this, there has been the development of WarnSim, which is an effective and flexible tool when using public safety networks. The tool is developed using Microsoft Visual C#. NET. The Windows application supports its operating platform while having a graphical user interface. Simulations with WarnSim can be performed in five steps, which include setup network topology, set up traffic sources, configure simulation parameters, run the simulation, and analysis of simulation results. Figures 14.3 and 14.4 show the WarnSim traffic trace generation and WarnSim simulation result or blocked calls, respectively.

It has been known that WarnSim is capable of supporting the functionality of any typical simulator. It has seven modules that include network module, call traffic module, call admission control module, a simulation configuration module, simulation configuration module, simulation process module, simulation statistics module, and a random variable module.

➢ Network module. The work of the network module is to model the systems. In addition, its other functions include closely monitoring the status of the systems, distributing calls to the covered system, and periodically updating the channel status in every system.
➢ Call traffic module. The function of this module is to prepare call traffic for simulation. Its two basic functions include generating call traffic based on user-defined distributions or importing traffic trace files from text files/databases.
➢ Call admission control module. This module is designed to communicate with the call traffic module and the networks module to assist in the determination of the availability of channels for a call to be established. Furthermore, call admission is important in managing and retrying mechanisms of blocked calls.
➢ Simulation configuration module. The work of this module is to track and closely monitor parameters such as simulation duration and interval.

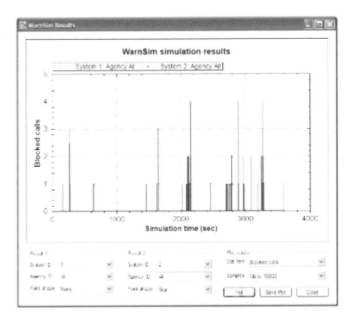

Figure 14.4 WarnSim simulation result (blocked calls).

➢ Simulation process module. This module has a timer that helps in controlling and synchronizing the operation of the WarnSim modules.
➢ Simulation statistics module. The function of this module is to collect real-time and summary statistics from the other modules. Moreover, its other functions include displaying and visualizing simulation results.
➢ Random variable module. This module generates random numbers and random variables. According to Matsumoto and Nishimura (1998), the module uses an MT random number generator, which is suitable for stochastic simulations. The variables that are generated are based on uniform, exponential, gamma, normal, log-normal, log-logistic, and Weibull distributions.

Figure 14.5 shows a schematic diagram of the call flow mechanism of WarnSim.

14.2.5 The Use Cases and Topologies of Public Safety Networks

The users and responders of public safety networks are often faced with a wide range of operational conditions and missions. Hence, as a way to effectively address these issues, there is an enormous need to dwell on sufficient voice and data communication services. Even though voice services are being used in tactical communication systems such as TETRA and P25, there is still an absence of a solid technology in the background to offer sufficient data services. When it comes to standard nominal conditions, a nationwide broadband wireless PS network is heavily reliant on wired networks that support fixed wireless BSs (Favraud and Nikaein 2015). The function of these BSs is to give planned coverage and give services to mobile entities such as hand-held user equipment or vehicle integrated devices that rely on seamless access to the core network (Apostolaras et al. 2015). It is important to note that the most important requirements for networks are

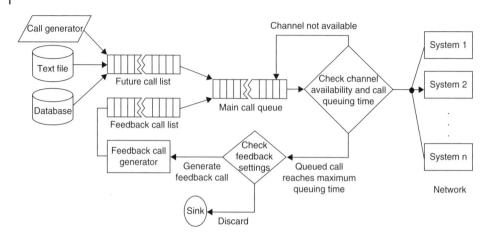

Figure 14.5 High-level diagram of WarnSim simulator (semanticscholars.org).

that they should be robust, reliable, and non-prone to malfunctions and outages. This is because when these networks have such qualities, they are able to withstand unexpected events such as earthquakes, tidal waves, and wildfires.

Figure 14.6 shows six different network topologies for public safety and the corresponding possible cases that the use cases and users may encounter based on the operational situation. The six topologies outlined are based on four criteria which are:

i. Availability of the backhaul link, which is the access to the core network from the BS
ii. BS interconnections
iii. BS mobility
iv. BS availability (UEs on-or off-network).

In the nominal case, which is case 1 in the figure, the structure of the BS is fixed and they have an advantage based on planned coverage because they receive complete service support. In addition, they experience full access to the core network as well as remote public safety services without any intermissions. Some of these intermissions include continuous link connectivity with the operation center, monitoring, and billing. Therefore, it is possible for a network to offer nominal access to PS UEs in the majority of operations such as emergency services, fire intervention, and law enforcement, which happens in covered cities and suburban scenarios (Barceló and Jordán 2000). This is also possible in areas where network deployment had been previously designed and the services provided covered a large extension.

Secondly, in the case of backhaul link failure due to faulty equipment, the core networks may become inaccessible to the fixed BS as seen in cases 2 and 3. However, it is important to note that this will be only dependent on the type of the position failure or the availability of the backup solutions, such as the one seen in the satellite backhaul in case 3. This is because the BSs may still have a sufficient interconnectivity to each other. Moreover, it is a good thing that portable BSs are still open to exploitation as they provide coverage on site where fixed BSs are not fully functional or may be experiencing faulty operations, as seen in case 4. Moving BSs also offers a dynamic platform and they can be used in fast moving forest fires, or in vehicular communication on land or at sea, which are challenging scenarios to plan inter-BS links (Bolotin 2013). Furthermore,

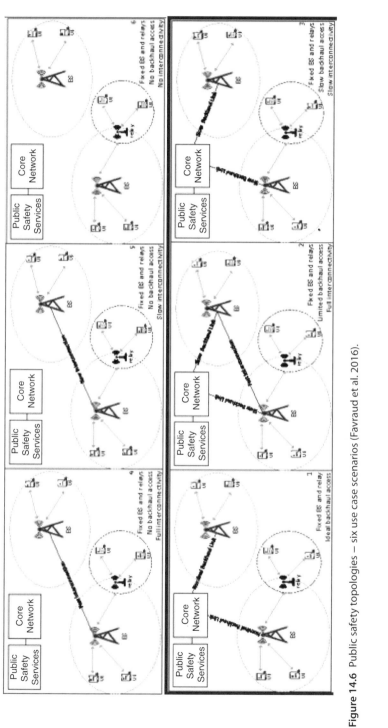

Figure 14.6 Public safety topologies – six use case scenarios (Favraud et al. 2016).

when using these portable or moving BSs, it can be challenging or close to impossible to maintain effective connectivity using the macro core network such as in cases 3, 4, and 5, as shown in Figure 14.6.

Lastly, when it comes to understanding the nature of mobility in these scenarios, uses in case 2, 3, and 4 are likely to come out of the coverage servicing area provided by the BSs or in time service provisioning due to intense mobility, as seen in case 6. Therefore, these inherent limitations, which include access to the core network, inter-node connectivity, BSs, and UEs mobility, may lack sufficient ability to provide the same quality in services offered to the users (Fang and Chlamtac 1999). For example, it would not be surprising to find that the billing and monitoring services are unavailable in some cases. Additionally, it is important to note that no matter the case, public safety network users must be able to utilize vital services such as voice and data group communications in every aspect regardless of the network topology in play. Hence, it can be deduced that public safety networks cannot be solely reliant on a planned network or fixed BSs.

The following are examples of public safety use cases.

Case 1: planned network with fixed BSs deployment and backhaul connectivity. Case 2: planned network with fixed BSs deployment and limited backhaul connectivity. Case 3: network with fixed BSs deployment and moving cells with limited backhaul connectivity assisted by satellite links, proximity services, and device-to-device communications. Case 4: no backhaul access in an unplanned network deployment of portable BSs. Case 5: moving cells in an unplanned network deployment. Case 6: missing BS coverage and proximity services.

14.2.6 Standard Developments in Public Safety Networks

The growing interest in public safety networks amongst many agencies in the world, especially when it comes to LTE, has motivated 3GPP to deal with this area and upgrade the LTE specifications. It is evident that after the establishments of the First Responder Network Authority (FirstNet), the United States engaged in various standardization activities. Figure 14.7 shows the first work dedicated to public-safety that was launched in 3GPP Release 11 alongside the creation of high power devices operating in Band 14 used in the United States and Canada to extend the coverage area (Gomez et al. 2014). Ever since, there have been several studies and works that shaped and motivated Releases 12 and 13, which are helpful in addressing particular and specific requirements of broadband public safety networks. Some of these requirements are listed below.

> ➢ Guaranteed access, which means that public safety networks should be readily available and accessible at any given time
> ➢ Quality of service (QoS), which guarantees that priority is given to critical calls.
> ➢ Reliability. This is the notion that public safety networks should provide services without any interruptions when online.
> ➢ Resiliency. This means that public safety networks should be adjustable to the ever-evolving technology advancements and swiftly change their operational requirements if need be.
> ➢ Roaming. In these kinds of networks, EUs should be able to effectively use and deploy public safety networks while also taking into consideration the commercial networks in case the former is unavailable.

Features

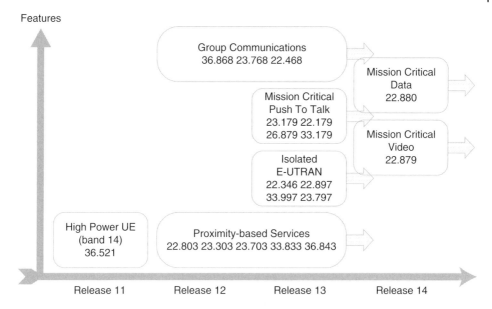

Figure 14.7 3GPP public safety oriented work items (Favraud et al. 2016).

➢ Spectrum efficiency, capacity, and coverage. There should be enough spectrum, which will, in turn, be significant in providing capacity, and coverage.
➢ Talk around/simplex. This means that users should be able to seamlessly communicate even in the case of broadband network unavailability, faultiness, or disruption.

The way in which public safety networks are gaining popularity through the LTE networks in the world has been heavily reliant on its architecture as it provides packet-based network services. It is significant to note that these services are independent from the underlying transport-related technologies (Favraud and Nikaein 2015). One of the core characteristics of LTE architecture is the solid dependency of the deployed BS, also known as eNB on the worldwide packet core network (EPC) for all types of their services offered by the UEs. However, it is crucial to point out that that this particular future inhibits UEs from effective communication when an eNB gets disconnected from the EPC. This is because the eNB service to the UEs will be interrupted even for local communications. As a solution to deal with the aforementioned challenge, 3GPP launched two series of work items that included device-to-device communication for enabling "proximity based services" (ProSe), and the continuity of service for PS EUs by the RAN and eNBs, especially in the case of backhaul failure for enabling operation on "isolated E-UTRAN" (Yuan 2012).

As defined in the 3GPP technical specification (TS) 22.346, isolated E-UTRAN has an objective of restoring the service of an eNB or a set of interconnected eNBs without dealing with the backhaul connectivity. Furthermore, the motivation for isolated E-UTRAN for Public Safety (IOPS) is to ensure there is the maintenance of the maximum level of communication for public safety users when the eNB connectivity to the EPC has become unavailable or is experiencing system disruption (Law and Kelton 2000). Normally, isolated E-UTRAN is capable of occurring on top of nomadic eNBs (portable BS, c.f. Ts 23. 797) deployments or on top of fixed eNBs that are experiencing system failure.

Isolated E-UTRAN should be able to support voice and data communications, MCPTT, ProSe, and group communications for public safety UEs, under coverage. In addition, they should be able to support mobility between BSs of the isolated E-UTRAN while at the same time watching out for security measures.

14.2.7 The Future Challenges in Public Safety

Because of the wide variety of applications associated with public safety, these types of communications have the need to offer huge flexibility and resilience (Matsumoto and Nishimura 1998). Therefore, it is very important for public safety communications to have the ability to adapt to various circumstances and mobility scenarios, which may be characterized by communication disruptions such as a damaged S1 interface or very limited EOC network access. Even though there has been increasing interest in the development of public safety solutions for isolated E-UTRAN, there are still challenges, which will be discussed below.

14.2.7.1 Moving Cells and Network Mobility

During an emergency or tactical scenario, it is crucial for field communications to be efficiently mobile and swiftly deployable to ensure great network coverage on the scene. In the current world, the status of the E-UTRAN has been regarded to be fixed and unspecified. For example, when high mobility is experienced, the challenge becomes network availability as the link connections to the EPC servers are dropped. Moreover, according to Apostolaras et al. (2015), the limited coverage of the moving cells as compared to the fixed eNBs enables inter-cell discovery features. This is because it is important to maintain proximity awareness, which is a crucial tool for network intelligence for self-healing. It means that the eNBs must be able to search for other eNBs directly or through the help of UEs. These UEs have extended abilities that are useful in interconnecting the eNBs, and ultimately, there will be synchronization of the access to the network.

14.2.7.2 Device-to-Device (D2D) Discovery and Communications

Assuming there is no network coverage, also known as the off-network scenario, public safety UEs need to discover and communicate with each other by partially taking control of the network's functionality (Liebhart et al. 2015). It is the work of the UEs to provide network assistance whenever the infrastructure nodes, which are the eNBs, are missing in action during terminal mobility or any sort of outage or malfunction. During such situations, UEs are often upgraded to deal with time synchronization reference, authentication, detection, network discovery, as well as attachment functions among others. Furthermore, UEs may be faced with the possibility of requesting the identity of neighboring UEs belonging to different public safety domains. This will require over-the-air sensing and self-reconfiguration functionality on the UE side. Nevertheless, it has been established that the most challenging factor for the public safety network is the support of stored data relaying from isolated neighboring UEs. This will be either UE to UE relay or UE to the network when they are in coverage.

14.2.7.3 Programmability and Flexibility

The notion of programmability and flexibility in future public safety systems has been seen to rapidly establish complex and mission-critical services with particular requirements in relation to service quality. This means the creation of an advanced degree of programmable network components capable of offering scalable and resilient network deployment on-the-fly without the need of previous network planning by using network function virtualization and software-defined networking (Huang et al. 2012). As a result, there will be the availability of open network interfaces, virtualization of networking infrastructure, and quick establishment of deployment of network services that possess the flexibility and an intelligent control and coordination framework. In line with the management of the entire life cycle of the public safety network, ensuring control and coordination of the framework is of importance. This can be a challenging task because it has to optimize resource allocation across several eNBs, because of the management of network topology, especially during splits and merges.

14.2.7.4 Traffic Steering and Scheduling

According to Apostolaras et al. (2015), going for a subset of eNBs to steer the data-plane traffic enables the users to get connected to the best-fitted networks based on the requirements of the QoS and network resource availability. While working toward the optimization of the network, traffic steering techniques can be leveraged to balance the network load and satisfy carrier and user demands.

14.2.7.5 Optimization of Performance Metrics to Support Sufficient QoS

It is known that public safety networks require the provision of sufficient services when the current eNB experiences backhaul connectivity or interruption. Besides using an isolated E-UTRAN operation such as exploiting the inter eNB connectivity, public safety networks also require mechanisms that are able to invoke the appropriate complementary resources, for example, additional bandwidth, alternate communication links, and complementary barriers. To achieve a more operational network, it is necessary that the same mechanisms make decisions based not only dwelling on the availability of complementary resources but also the indicators and characteristics of communication performance such as latency, throughput, and spectral efficiency, amongst others.

14.2.8 Toward a Convergence Future of Public Safety Networks

In the existing LTE architectures, eNBs stand out as being the active elements that are responsible for the management and control of the RAN. By contrast, UEs are perceived to be the passive clients based on the eNBs perspective in terms of obeying its own rules and policies. Hence, it can be deduced that the eNBs and UEs relationship fits that of a master–slave communication model that is very crucial in meeting the requirements for fixed network topology. Ideally, network mobility will continue to gain interest, and the subject is becoming more and more attractive. Based on the previous discussion, a converged future for public safety networks will involve the in-depth utilization of eNBs and UEs.

While focusing on this approach, researchers have established two significant solutions that enhance inter-eNB link connectivity and the restoration of a disrupted air interface. These solutions have obviously been made possible by the use of advanced UEs (eUEs) and eNBs (e2NBs). In a more specific way, the first one refers to a UE-centric network-assisted solution. This means that the eUEs possess the advanced capabilities of linking multiple UE stacks, which result in interconnections between the adjacent eNBs. On the other hand, the second solution is the network-centric solution (Hu et al. 2015).

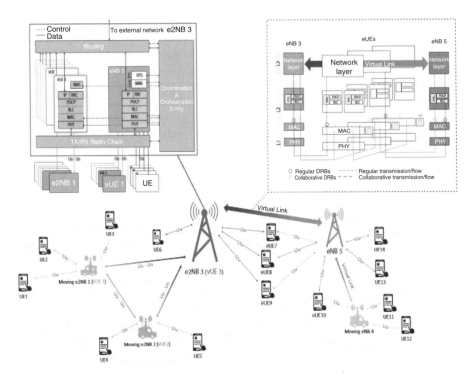

Figure 14.8 Solution enhancing inter-eNB link connectivity. Source: UE and e2NB architecture for public safety (Favraud et al. 2016).

Essentially, this means that the eNB stack gets extended with other several UE stacks thereby resulting in a e2NB. The topology of these two solutions is presented diagrammatically in Figure 14.8 together with an illustration of the eUE and e2NB architectures.

Meshing of isolated or moving eNBs is enabled either:

(i) by leveraging eUEs as intermediate packet forwarders (UE-centric), thus creating virtual links between eNBs, or
(ii) by leveraging e2NBs functionality of encompassing multiple vUEs (network-centric), thus restoring disrupted eNB–eNB communication.

14.3 Conclusion

In conclusion, the future convergence of networks will depend on LTE, as specified by the 3GPP, as a technology that is becoming highly relatable to 4G cellular networks. This is because the majority of operators and agencies have adopted it throughout the world. Since it is a technology that is being absorbed by the majority of technology companies, LTE has faced various challenges over the past few years. These include factors such as cellular networks, capacity crunch, ultra-high bandwidth, and ultra-low latency, a massive number of connections, very fast mobility, and diverse spectrum access that fuels the pace toward a 5G scenario. Furthermore, it has been established that LTE is the solution to future 5G networks, which will eventually play a role in public safety communications. In the contemporary world, the need for reliable telecommunication is constantly increasing. Since various factors call for mission-critical communications, the nature of public safety communication is heavily reliant on conventional LMR systems. These radio systems are a good source of field communication, especially when in liaison with the command and control centers. However, it is important to note that various issues of incompatibility arise because of the constant evolution of the LMR systems. Therefore, public safety networks are highly susceptible to incompatible systems. Such outcomes within public safety communities have prompted them to devise better ways to deal with communication problems. Moreover, the current public safety agencies utilize a combination of low bandwidth and high-speed broadband data. The latter is mainly used in large commercial networks because of its capability to support enormous response efforts and functions such as text messaging, license plate queries, and digital dispatch.

References

Apostolaras, A., Nikaein, N., Knopp, R. et al. (2015). Evolved user equipment for collaborative wireless backhauling in next-generation cellular networks. In: *2015 12th Annual IEEE International Conference on Sensing, Communication, and Networking (SECON)*, 408–416. IEEE.

Barceló, F. and Jordán, J. (2000). Channel holding time distribution in public telephony systems (PAMR and PCS). *IEEE Transactions on Vehicular Technology* 49 (5): 1615–1625.

Bolotin, V. (2013). Telephone circuit holding time distributions. In: *Proceedings of the ITC*, vol. 14, 125–134.

Carla, L., Fantacci, R., Gei, F. et al. (2016). LTE enhancements for public safety and security communications to support group multimedia communications. *IEEE Network* 30 (1): 80–85.

Casoni, M., Grazia, C.A., Klapez, M. et al. (2015). Integration of satellite and LTE for disaster recovery. *IEEE Communications Magazine* 53 (3): 47–53.

Department of Homeland Security. (2015). Interoperability Continuum. Department of Homeland Security. https://www.dhs.gov/sites/default/files/publications/interoperability_continuum_brochure_2.pdf

Desourdis, R.I., Smith, D.R., Speights, W.D. et al. (2013). *Emerging Public Safety Wireless Communication Systems*. Washington: Artech House.

Desourdis, R.I. Jr., Dew, R.A., O'Brien, M., and Hinsch, H. (2015). *Building the FirstNet Public Safety Broadband Network*. Norwood: Artech House.

Fang, Y. and Chlamtac, I. (1999). Teletraffic analysis and mobility modelling of PCS networks. *IEEE Transactions on Communications* 47 (7): 1062–1072.

Favraud, R. and Nikaein, N. (2015). Wireless mesh backhauling for LTE/LTE-A networks. In: *MILCOM 2015–2015 IEEE Military Communications Conference*, 695–700. IEEE.

Favraud, R., Apostolaras, A., Nikaein, N., and Korakis, T. (2016). Toward moving public safety networks. *IEEE Communications Magazine* 54 (3): 14–20.

Ferrús, R. and Sallent, O. (2015). *Mobile Broadband Communications for Public Safety: The Road Ahead Through LTE Technology*. New York: Wiley.

Frost, V.S. and Melamed, B. (1994). Traffic modelling for telecommunications networks. *IEEE Communications Magazine* 32 (3): 70–81.

Gomez, K., Goratti, L., Rasheed, T., and Reynaud, L. (2014). Enabling disaster-resilient 4G mobile communication networks. *IEEE Communications Magazine* 52 (12): 66–73.

Hu, Y. C., Patel, M., Sabella, D., Sprecher, N., & Young, V. (2015). Mobile edge computing—A key technology towards 5G, *ETSI white paper* #11. 1–16. ETSI

Huang, J., Qian, F., Gerber, A. et al. (2012). A close examination of the performance and power characteristics of 4G LTE networks. In: *Proceedings of the 10th International Conference on Mobile Systems, Applications, and Services*, 225–238. ACM.

Law, A.M. and Kelton, W.D. (2000). *Simulation Modelling and Analysis*, vol. 3. New York: McGraw-Hill.

Liebhart, R., Chandramouli, D., Wong, C., and Merkel, J. (2015). *LTE for Public Safety*. Wiley.

Matsumoto, M. and Nishimura, T. (1998). Mersenne Twister: a 623-dimensionally equidistributed uniform pseudo-random number generator. *ACM Transactions on Modeling and Computer Simulation (TOMACS)* 8 (1): 3–30.

Nikaein, N., Knopp, R., Kaltenberger, F. et al. (2014). OpenAirInterface: an open LTE network in a PC. In: *Proceedings of the 20th Annual International Conference on Mobile Computing and Networking*, 305–308. ACM.

Rajaratnam, M. and Takawira, F. (2001). Handoff traffic characterization in cellular networks under nonclassical arrivals and service time distributions. *IEEE Transactions on Vehicular Technology* 50 (4): 954–970.

Sharp, I. (2014). *Delivering Public Safety Communications with LTE*. 3GPP.

Sharp, D.S., Cackov, N., Laskovic, N. et al. (2004). Analysis of public safety traffic on trunked land mobile radio systems. *IEEE Journal on selected areas in Communications* 22 (7): 1197–1205.

J. Song, Ljiljana Trajkovic, (2005) Modeling and performance analysis of public safety wireless networks, Conference: 24th IEEE International Performance, Computing, and Communications Conference, 2005. IPCCC 2005.

Weny News (2019). *Public Safety LTE Market 2019 Size, Share, Recent Trends, Industry Growth, Historical Analysis, Future Demand, Opportunities, Regional and Global Forecast to 2025*. Weny News http://www.weny.com/story/40041925/public-safety-lte-market-2019-size-share-recent-trends-industry-growth-historical-analysis-future-demand-opportunities-regional-and-global-forecast.

Yuan, Y. (2012). *LTE-Advanced Relay Technology and Standardization*. Springer Science & Business Media.

Index